Aluminum Alloys:
Fabrication, Characterization
and Applications II

T0334416

ALUMINUM ALLOYS: FABRICATION, CHARACTERIZATION AND APPLICATIONS II

Related titles include:

- *Aluminum Alloys: Fabrication, Characterization and Applications (2008)*

- *Aluminum Alloys for Transportation, Packaging, Aerospace, and Other Applications (2007)*

- *Shape Casting of Metals II (2007)*

HOW TO ORDER PUBLICATIONS

For a complete listing of TMS publications, contact TMS at (724) 776-9000 or (800) 759-4TMS or visit the TMS Knowledge Resource Center at http://knowledge.tms.org:

- Purchase publications conveniently online and download electronic publications instantly.
- View complete descriptions and tables of contents.
- Find award-winning landmark papers and webcasts.

MEMBER DISCOUNTS

TMS members receive a 30 percent discount on TMS publications. In addition, members receive a free subscription to the monthly technical journal *JOM* (both in print and online), free downloads from the Materials Technology@TMS digital resource center (www. materialstechnology.org), discounts on meeting registrations, and additional online resources to name a few of the benefits. To begin saving immediately on TMS publications, complete a membership application when placing your order or contact TMS:

Telephone: (724) 776-9000 / (800) 759-4TMS

E-mail: membership@tms.org or publications@tms.org

Web: www.tms.org

Aluminum Alloys:
Fabrication, Characterization and Applications II

Proceedings of symposia sponsored by the Light Metals Division
of The Minerals, Metals & Materials Society (TMS)

Held during TMS 2009 Annual Meeting & Exhibition
San Francisco, California, USA
February 15-19, 2009

Edited by

Weimin Yin
Subodh K. Das
Zhengdong Long

A Publication of

A Publication of **The Minerals, Metals & Materials Society (TMS)**
184 Thorn Hill Road
Warrendale, Pennsylvania 15086-7528
(724) 776-9000

Visit the TMS Web site at
http://www.tms.org

Statements of fact and opinion are the responsibility of the authors alone and do not imply an opinion on the part of the officers, staff, or members of TMS. TMS assumes no responsibility for the statements and opinions advanced by the contributors to its publications or by the speakers at its programs. Registered names and trademarks, etc., used in this publication, even without specific indication thereof, are not to be considered unprotected by the law.

No part of this book may be reproduced, stored in a retrieval system, or transmitted in any form or by any means, electronic, mechanical, photocopying, microfilming, recording, or otherwise, without written permission from the publisher.

ISBN Number 978-0-87339-735-3

If you are interested in purchasing a copy of this book, or if you would like to receive the latest TMS Knowledge Resource Center catalog, please telephone (724) 776-9000, ext. 270, or (800) 759-4TMS.

TABLE OF CONTENTS
Alumiunum Alloys: Fabrication, Characterization and Applications II

Alumiunum Alloys: Fabrication, Characterization and Applications II

Development and Application

Processing and Properties

Formability and Texture

Materials Characterization

Modeling and Corrosion

Composite and Foam

Preface

This publication comprises the proceedings of the symposium "Aluminum Alloys: Fabrication, Characterization and Applications" in the TMS 2009 Annual Meeting & Exhibition held in San Francisco, California, United States. The symposium successfully received high quality presentations and papers from research universities, national laboratories, and downstream processing industries, which indicates the continued strong interests in research and development of aluminum alloys. This collection of contributed papers is aimed to provide insights and updates on fundamental research, product development and implementation for the broad applications including transportation (automotive, aerospace and marine), packaging and other key product segments. Besides the traditional focuses, attention has been paid particularly on computational modeling and composite due to their increasing importance in both academic research and industry applications. Aligning with the presentation sessions, the papers published are divided into six sections.

The editors would like to thank all authors for submitting their work and all TMS staff members who contributed so much to the success of the symposium and the proceedings. Thanks are also due to the Light Metals Division of TMS for sponsoring this symposium.

Weimin Yin
Williams Advanced Materials Inc.
Brewster, New York

Subodh K. Das
Phinix, LLC
Lexington, Kentucky

Zhengdong Long
Kaiser Aluminum Corporation
Spokane, Washington

About the Editors

Weimin Yin is a principal R&D scientist at Williams Advanced Materials Inc. in Brewster, New York. He has been active for the past two decades in research and development of innovative technologies and products on a wide range of materials including high temperature materials for aerospace applications, aluminum alloys for packaging industries and sputtering targets for thin film magnetic recording. Dr. Yin has extensive industrial and academic working experience with expertise in processing techniques, microstructure characterization and mechanical testing. His current research interests are in the development of industrial processes for manufacturing magnetic thin film products through a broad range of melting, thermo-mechanical technologies and powder metallurgy. He has two patents to his credit and more than 30 published papers. He received his doctorate in materials science from Polytechnic University in Brooklyn, New York, in 2003.

Dr. Das is the CEO & Founder of Phinix, LLC. Phinix, LLC, based in Lexington, Kentucky, USA, is dedicated to the promotion, development and implementation of low carbon footprint manufacturing technologies and carbon management and trading for the Minerals, Metals and Materials Industries to help Globally Responsible Resource Management.

Dr. Das has over 30 years of global experience in manufacturing and technology areas. Dr. Das is well recognized and respected expert and consultant to the aluminium industry specializing in the areas of technology, recycling and new product and process developments.

 **Zhengdong Long** is an Alloy Development Engineer at Kaiser Aluminum Corporation located in Spokane, Washington, United States. Prior his current position, he was Associate Director at Center for Aluminum Technology, University of Kentucky. He was Senior Engineer at Central Iron and Steel Research Institute in Beijing, China before his last career. He has been active for over twenty years in the areas of physical and mechanical metallurgies of aluminum alloy and Superalloys. Dr. Long's diverse experience includes new alloy and process development, product and process improvement, product root cause failure analysis, and new materials application. He has been focused on the research of thermal-chemistry, microstructure and texture development during casting, thermo-mechanical processing, heat treatment processes and their effects on mechanical and corrosion properties. He specializes in microstructure characterization, mechanical property testing, process modeling to simulate casting, rolling, and forging processes, thermodynamic modeling to simulate solidification, phase diagrams and transformation. Dr. Long hold one patent to his credit and more than 20 published papers. He received his doctorate in Materials Science and Engineering from Central Iron and Steel Research Institute in Beijing, China, in 2000, and MBA from University of Kentucky in 2007.

Aluminum Alloys:
Fabrication, Characterization and Applications II

Development and Application

Session Chair

Shridas Ningileri

Aluminum Alloys: Fabrication, Characterization and Applications II
Edited by: Weimin Yin, Subodh K. Das, and Zhengdong Long
TMS (The Minerals, Metals & Materials Society), 2009

DEVELOPMENT OF LOW COST, HIGH PERFORMANCE AlZn4.5Mg1 ALLOY 7020

ARL[1], John Chinella[2],

[1]U.S. Army Research Laboratory
[1,2]AMSRD-ARL-WM-MD, Aberdeen Proving Ground, MD 21005, USA

Keywords: Aluminum, Armor, Al 7020, Regression Analysis

Abstract

This paper describes properties, processing, and performance of Al-Zn-Mg alloy 7020 for armor and commercial applications. Comparisons are made with alternative aluminum armors' chemistry, properties, levels of strength and ductility, weld properties, and resistance to stress and exfoliation corrosion. The advantages for development and optimization of alloy 7020 for vehicle armor or welded structures are identified to be: (1) low thermal sensitivity of the microstructure and mechanical properties to deleterious effects from reheat or solution-treatment and air-quench, (2) low Zn-Mg alloy and production costs, (3) adequate thick-plate strength and ductility for ballistic protection, and (4) high levels of weld strength and ductility in the natural or artificial aged condition. Concerns include: (1) optimizing thickness-dependent combinations of strength, toughness, and armor-threat-protection; and (2) optimizing processing and manufacturing techniques for mechanical properties and durability against stress and exfoliation corrosion.

Introduction

Humanitarian and military missions require logistical and combat vehicles capable of operation in largely undeveloped areas, on and off-road in areas of rough terrain. The hazards of these operations include the low-cost denial of area weapons, the landmine and improvised explosive device (IED) [1-3].

Effective protection for cargo transport, patrol vehicles, armored personnel carriers, and combat vehicles were developed in South Africa against mine blast threats and IEDs [1,2]. The light vehicle 4-6 passenger Cougar, retrofit on Land-Rover or Nissan chassis, was claimed to be the best land-mine protected vehicle of the Rhodesian war [1]. The heavy vehicle and 10-passenger capacity Buffel, built on a Unimog chassis, or the new monocoque-body 12-passenger Casspir proved among the most successful designs.

Casualty statistics [1, 2] proved that successful vehicle design elements should provide for long wheel bases, thin and tough ductile blast shields that deform but that do not become detached or displaced into the vehicle, high standoff height Vee-shaped hulls of ductile and tough mild steel that dissipate and deflect blast, hardened seating and safety belts, roll bars, over-pressure blow-off roof panels and fuel tanks.

Threats from mine and IED explosions include fragmentation, shock, overpressure, heat and combustion, and forces of acceleration and deceleration. The threats of mine casing fragments are similar to those of artillery. The ballistic protection, floor, and vehicle structure must resist deformation and stresses of the blast to avoid being torn apart, to form secondary fragments, and to avoid openings of doors or windows. Rapid deflection of floors caused by shock loading may cause fractures of feet, ankles, and legs. Acceleration and deceleration of the vehicle (see Figure 1 [3]) may rapidly subject the crew to incapacitating and lethal injuries of the neck and spine. The criterion of NASA for jet ejection seat force (23 g > 7 ms, g = 9.8 m/s acceleration unit of earth gravity) has been used for estimation of back spinal injury [3].

Armor Materials for Protection against Blast and Projectiles

Materials used in the Cougar, Buffel, and Casspir Vee-hulls vehicles were "mild" steel resistant to the formation of fragments and splinters. These steels have advantages of high fracture toughness and ductility levels that resist crack initiation and propagation. Disadvantages of low-alloy steel include high density, and a temperature-sensitive yield stress [4] that increases the yield to tensile strength ratio and notch-sensitivity, lowering impact and fracture toughness and the ability for crack arrest under conditions of low temperatures and increased strain rate.

In comparison to steel [5] aluminum armor may provide weight savings to defeat the mine blast and small caliber armor piercing and fragment threats to mine resistant vehicles. With adequate combinations of strength, elongation, and fracture ductility (see Tables 1, 2), [6-14] aluminum armors, ranked 1 to 5 by yield strength (see Figure 2 [15]), provide excellent protection against fragment simulating projectiles (FSPs) beyond a critical thickness, shown by comparisons to V50 – areal density (AD) mean estimates (MEs) of 5083 and 7039 Al armors [9,13] and rolled homogeneous armor (RHA) steel [16], along with 95% confidence intervals (CIs) [17] for the 5083 MEs and single point future predictions (SPFPs). At equivalent weight, ballistic protection levels of aluminum versus armor piercing projectiles (AP) is largely dependent on yield strength and may be somewhat worse to better than average than RHA (see Figure 3 [15]). At low temperatures typical of aerospace (-88 °C) or cryogenic environments (-196 °C) strength, elongation, and K_{IC} fracture toughness of aluminum alloys typically increase over their room temperature values [18] suggesting improved resistance to ballistic perforation.

In addition to having the necessary properties of strength, elongation, and ductility for ballistics protection, candidate materials for mine resistant vehicles must resist structural failure by fracture initiation and propagation. Resistance to crack propagation by a full spectrum of deformation and fracture modes in rolled plates of aluminum alloys and under conditions of plane stress to plane strain has been characterized by fracture extension resistance (R-curve) properties determined by the dynamic tear (DT) test [19]. Use of aluminum materials to meet requirements for ballistic protection and structural integrity in mine-resistant vehicles will require careful selection of material properties to resist projectile and fragment threats and crack initiation and propagation. Vehicle design schemes to optimize protection levels include aluminum hulls, with aluminum of varied strengths, elongation, and toughness, and add-on or standoff armors to optimize protection from mine blast, fragments, or projectiles [20].

Low strength grades of the 5XXX series of aluminum have been shown to provide high resistance to crack extension and ability for fracture arrest [19]. Disadvantages of the 5XXX alloys are: (1) low levels of ballistic protection against AP projectiles (see Figure 3); (2) long-term exposure to elevated temperature in alloys with Mg content greater than 3% degrades resistance to stress corrosion [21]; (3) a typical loss of 50% of yield strength in fusion weld metal [22-25], (4) for thick plate the difficulty for cold work strengthening results in low strength and less efficient protection per weight [9, 11].

Aluminum 7020 Armor: Advantages and Concerns

The medium strength, Cu-free, low Mg-content alloy (<1.4 Mg) 7XXX Al-Zn-Mg alloys [24] including 7020-T651 (see Tables 3-4 and Figures 4-5) and 7005-T651 [18] (see Tables 1-2) have strength and ductility levels that typically provide improved resistance to penetration by fragments and armor piercing projectiles in comparison to 5083 aluminum and RHA (see Figures 2-3). With relatively high strength, the fracture arrest capability levels of these alloys as shown by 7005-T651 [19] may be of concern for thick sections. For high levels of protection versus projectiles, strength and ductility levels of alloy 7020-T651 can be maintained over 50 mm-

4

thickness [11] (see Figure 4 and Table 4).

During continuous heating, strengthening precipitates in Al-4.5Zn-1.2Mg alloy either completely dissolve above the solvus temperature of 350°C or at lower temperatures undergo reversion, a quick decrease of precipitate volume fraction, and with times >100 seconds, a progressive slow coarsening stage that increases the volume fraction and precipitate size [6,26]. For vehicle-structural integrity, fusion welds of the Cu-free Al-Zn-Mg alloys typically regain 70% of their yield strength after 30 days of natural aging, and with artificial aging 100% of their original yield strength [23, 27] (see Table 5).

High levels of Zn makes aluminum solid solutions of Cu-free 7XXX alloys more electrochemically active and susceptible to galvanic corrosion [21]. The highest level of durability to SCC for 7020 is achieved by the T7 temper [10] (see Table 6). SCC resistance in Al-Zn-Mg alloys is achieved together by alloy chemistry and processing, product design, and manufacturing practice [6, 21, 24]. Following fusion welding, resistance to SCC [27] and exfoliation [28] is achieved by artificial aging (see Table 7).

The Cu-free 7XXX aluminum alloys contain low cost Zn and Mg alloy elements, and the alloys with Mg contents < 1.4 percent have low quench sensitivity that allows slow cooling rates and simple and economical heat treat processing by air-cooling. Air-cooling in these Al-Zn-Mg alloys provides optimal microstructures for improved resistance to stress corrosion cracking [6, 24, 27]. The alloy microstructures [26] obtained from duplex heat treatment to T7 tempers in 7XXX alloys provide improved levels of fracture toughness [6, 24, 27, 30], fracture ductility (see Figure 5), and protection versus FSP threats (see Figure 2).

Development Objectives for 7020

Objectives are to develop confidence in processing thin to thick plate 7020 aluminum armor for superior levels of weld strength and ductility, environmental durability, and ballistic protection to levels equal or superior to 5083 and 5059 aluminum armors. The approach shall include the determination of microstructure, properties, and performance of commercial plate in the T651 condition, and T6 and T7 conditions following MIG (metal inert gas) fusion welding and or artificial aging. Investigation of the MIG weld process shall include selections of filler metal alloy, travel speed and power. The quality, improvements, and performance of the parent metal and weldments shall be determined by microstructure, mechanical characterization, V50 performance and ballistic shock resistance, and determination of K_{ISCC} or critical levels of stress and time for the initiation of stress corrosion cracking. To obtain high levels of toughness similar to that for 7075-T7XX tempers (see Figure 5) and [30]), slow air cooling and overage or duplex high temperature temper treatments will be investigated for achievement of improved fracture ductility, crack arrest capability, and protection versus fragment threats and FSP projectiles.

5

Tables

Table 1. Aluminum armor and commercial alloy mechanical properties.

Alloy	0.2% Y.S. (MPa)	U.T.S. (MPa)	El. (%)	Data	Reference
7005-T651	290 (42)	370 (54)	15	typical	[6]
6061-T651	300 (44)	337 (49)	19	experiment	[15]
7018-T6751	300 (44)	360 (52)	12	typical	[7]
5083-H131	319 (46)	377 (55)	9.3	experiment	[8]
5059-H131	290 (42)	345 (50)	8	> 50.8mm, min.	[9]
5059-H131	296 (43)	393 (57)	7	12.7-50.8 mm, min.	[9]
7020-T7651	318 (46)	-	-	experiment	[10]
7020-T651	347 (50)	397 (58)	13	60.0 mm, cert	[11]
7020-T651	351 (51)	401 (58)	14	25.4 mm, cert.	[11]
2519-T87	423 (61)	465 (67)	12.4	experiment	[8, 12]
7039-T64	400 (58)	458 (66)	13.6	experiment	[8, 13]

Table 2. Aluminum Association chemical composition limits [14].

Alloy	Si	Fe	Cu	Mn	Mg	Cr	Ni	Zn	Ti	V	Zr	Date
6061	0.40 0.8	0.7	0.15 0.40	0.15	0.8 1.2	0.04 0.35	-	0.25	0.15	-	-	1954 USA
5083	0.40	0.40	0.10	0.40 1.0	4.0 4.9	0.05 0.25	-	0.25	0.15	-	-	1954 USA
5059	0.45	0.50	0.25	0.6 1.2	5.0 6.0	0.25	-	0.40 0.9	0.20	-	0.05 0.25	1999 Ger.
7005	0.35	0.40	0.10	0.2 0.70	1.0 1.8	0.06 0.20	-	4.0 5.0	0.01 0.06	-	0.08 0.20	1962 USA
7018	0.35	0.45	0.20	0.15 0.50	0.7 1.5	0.20	0.10	4.5 5.5	0.15	-	0.10 0.25	1978 UK
7020	0.35	0.40	0.20	0.05 0.50	1.0 1.4	0.10 0.35	-	4.0 5.0	Zr+Ti 0.25	-	0.08 0.20	1972 EAA
7039	0.30	0.40	0.10	0.10 0.40	2.3 3.3	0.15 0.25	-	3.5 4.5	0.1	-	-	1962 USA
2519	0.25	0.30	5.3 6.4	0.10 0.50	0.05 0.40	-	-	0.10	0.02 0.10	0.05 0.15	0.05 0.25	1985 USA
7017	0.35	0.45	0.20	0.05 0.50	2.0 3.0	0.35	0.10	4.0 5.2	0.15	-	0.10 0.25	1978 UK

Table 3. Certified chemical analysis of the commercial 7020-T651 alloys, weight-%, 14 plates among 12, 15, 20, 25, 30, 40, 50, 60, 80 mm-thick plates [11].

Chemistry	Si	Fe	Cu	Mn	Mg	Cr	Ni	Zn	Zr	Ti
Average	0.100	0.263	0.161	0.249	1.240	0.169	0.006	4.471	0.034	0.133
Std Dev	0.033	0.040	0.022	0.037	0.015	0.031	0.004	0.113	0.004	0.008

Table 4. Certified mechanical properties of
commercial 7020-T651 plates [11].

Thickness (t)	0.2 % Y.S.	U.T.S.	Elong.	Hardness
(mm)	(MPa)	(MPa)	(%)	(HB)
12	360	408	16.8	-
15	361	409	13.7	-
20	351	403	14.0	116
25	351	401	13.9	-
30	351	401	15.0	116
40	340	393	14.0	117
50	348	397	12.8	-
60	347	397	13.0	115
80	320	370	13.1	-
Aver.,	351	401	14	116
Std. Dev.	8	6	1	1

Average and standard deviations: plate ≤ 60 mm thick

Table 5. GMAW mechanical properties and strength efficiencies.

Alloy	0.2%Y.S.	U.T.S.	El.(%)	Filler	Condition	Eff. (%)	Eff. (%)	Reference
	(MPa, ksi)	(MPa, ksi)				Y.S.	U.T.S	
5083-H131	319 (46)	377 (55)	9.3		base metal	-	-	[8]
	152 (22)	283 (41)	12.2		as-welded	48	75	[22,23,25]]
2519-T87	423 (61)	465 (67)	12.4		base metal	-	-	[8,22,25]
	209 (30)	301 (44)	4.4	2319	as-welded	49	65	[22,25]
7005-T6	290 (42)	370 (54)	15		base metal	-	-	[6,7,22]
	204 (30)	322 (47)	11		weld, nat. age 3 mo	70	87	[6,22]
7020-T651	352 (51)	390 (57)	18.5	5356	base metal	-	-	[27]
7020-T6	356	399	12.8	5356	1-P weld+art age	100	100	[27]
7020-T6	366	410	8.9	5356	2-P weld+art age	100	100	[27]

[27] 7020 = 1.6 mm sheet

Table 6. SCC stress intensity data for
aluminum alloys, SL orientation, initiation.

Alloy	K_Q, K_{ISCC}	Environment	Ref.
2519-T87	21. 0	3.5% sol.	[8]
5083-H131	21. 3	3.5% sol.	[8]
5083-H131	10. 6	3.5% sol.	[8]
7017-T651	6. 7	3.5% sol.	[10]
7039-T64	6..3	3.5% sol.	[8]
7039-T64	4. 3	3.5% sol.	[8]
7020-T7651	28. 2	seawater	[10]
7018-T7651	38. 1	seawater	[10]
7017-T7651	6. 0	seawater	[10]

Note: method of reference [10] by crack propagation and arrest.

Table 7. Stress corrosion cracking, 5356 weld, mechanical test results.

Alloy/Filler	Test Condition	Stress (MPa)	Failures	Metallurgical Condition
7020-T6 / 5356	40 °C, 80% RH	200	0/3	1-P weld + art age
	40°C, 80% RH	160	2/3	2-P weld
	40 °C, 80% RH	225	0/5	sol. treat. air quench, art.-age
	40°C, 3% NaCl, pH4	225	0/3	480°C sol. treat. air quench, art.-age
	40°C, 3% NaCl, pH4	200	0	1-P weld + art. age, 1030 -1430 mV pot.
2519-T87 / 2319	3% NaCl, Al, C-ring, 90 hrs	138	4/4	Weld, bead, ST,

Notes: 7020 test data from [27]; 2519 test results from [8]; RH = relative humidity.

7

Figures

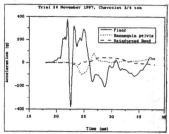

Figure 1. Time-acceleration effects, medium-sized blast mine equivalent to 7.5 kg TNT [3].

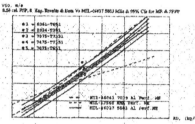

Figure 2. Commercial Al alloys protection V50 velocity by areal density (AD) versus 0.50 cal. FSP projectiles [15]. Comparisons to: 7039-T64 Al armor and rolled homogeneous armor (RHA) steel. V50 order 3, 4, 2, 5, 1.

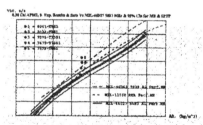

Figure 3. Commercial Al alloys protection V50 velocity level by areal density (AD) versus 0.30 cal.APM2 projectiles [15]. V50 order 5, 3, 4, 2, 1.

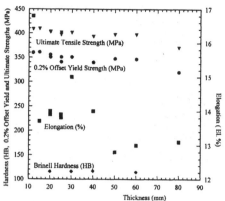

Figure 4. Mechanical properties and hardness of alloy 7020-T651 commercial plate, 12 mm to 80 mm-thick [11].

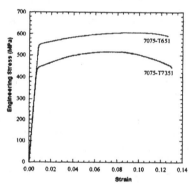

Figure 5. Engineering stress-strain plastic flow curves. Fracture ductility of T651 and T7351 temper [15].

References:

[1] Stiff, Peter. *Taming the Landmine*, Galago Publishing Ltd. Alberton, RSA; 1986.

[2] Lester, Craig. "Protection of Light Skinned Vehicles Against Landmines – A Review", Department of Defense, Defense Science and Technology Organization, June, 1996.

[3] L.P. LaFrance, "Mine Blast Protection System For Military Vehicles," PVP-Vol. 361, *Structures Under Extreme Loading Conditions-1998*, ASME 1998, 305-309.

[4] Hertzberg, R.W. *Deformation and Fracture Mechanics of Engineering Materials*, John Wiley & Sons, N.Y, 1989.

[5] Crouch, I.G. *Metallic Armor-From Cast Aluminum Alloys to High-strength Steels*, Materials Forum, 1988, 12, 31-37.

[6] Kent, K.O. "Weldable Al.Zn.Mg Alloys," *Metallurgical Reviews*, Vol. 15, Issue 147, 1970, pp. 135-146.

[7] Bayley, M. J. "Offshore Role for Lightweight Armour," *Structures Under Shock and Impact. II* Portsmouth, UK; 16-18 June 1992, pp 545-554, 1992.

[8] Chase, M., N. Kackley, and W. Bethoney. "Engineering and Ballistic Properties of a Newly Developed 2XXX Series Aluminum Alloy Armor," *The Proceedings of the 9th International Symposium on Ballistics*, Shrivenham, UK, 29 April -1 May 1986; Royal Military College of Science, Session VI, Pt. 2, pp. 511-521.

[9] MIL-DTL-46027K. Armor Plate, Aluminum Alloy, Weldable 5083, 5456, & 5009; 2007.

[10] Holroyd, N.J.H., and D. Hardie, "Effect of Inherent Defects on Initiation of Stress-Corrosion Cracks in Weldable Al-Zn-Mg Alloys," *Metals Technology*, June 1982, Vol. 9, pp. 229-234.

[11] Certification Test Reports: (a) Aleris Aluminum Koblenze Gmbh; (b) Alcan Aluminum Valais (Sierre); (c) Hoogovens Aluminum, Koblenze GmbH. 12-80 mm 7020-T651 plate.

[12] MIL-DTL-46192C *Detail Specification Aluminum Alloy Armor Plate (1/2 to 4 inches Thick), Weldable Alloy 2519* (1998).

[13] MIL-DTL-46063H. *Armor Plate, Al Alloy 7039.* 1998.

[14] Registration Record Series Teal Sheets, *International Alloy Designations and Chemical Composition Limits for Wrought Aluminum and Wrought Aluminum Alloys*, (www.eaa.net/en/about-aluminium/standards/international-**registration**), The Aluminum Association, Arlington, VA, April 2006.

[15] Chinella, J.F. and T. Jones. "Evaluation of Commercial Aluminum Alloys for Armor and Vehicle Upgrade," ARL-TR-4596, Sept. 2008, U.S. Army Research Laboratory, Aberdeen Proving Ground, MD 21005.

[16] MIL-A-12560H. *Armor Plate, Steel, Wrought, Homogeneous* (1990).

[17] Montgomery, D.C.; Peck, E.A. *Introduction to Linear Regression Analysis*, 2nd ed.; Wiley & Sons: New York, 1992.

[18] ASM Handbook, Vol. 2, Props. and Select.: *Nonferrous Alloys and Special-Purpose Materials*. ASM, Metals Park, OH, 1998, pp. 54-60, 92–93, 108–111.

[19] Judy, R.W. and R.J. Goode. "Dynamic Tear Tests of 3-In.-Thick Aluminum Alloys," ASTM STP 536, American Society for Testing and Materials, 1973, pp. 377-390.

[20] Ogorkiewicz, R.M., "Armor for Light Combat Vehicles," *Jane's International Defense Review*, July 01, 2002.

[21] Summerson, T.J., and D.O. Sprowls, "Corrosion Behavior of Aluminum Alloys," *Aluminum Alloys Their Physical and Mechanical Properties*, Vol. III, eds. E.A. Starke, Jr. and T.H. Sanders Jr., Engineering Materials Advisory Services Ltd. Warely, UK, 1986.

[22] Wolfe, T.D., and S.A. Gedeon, "Weldability of 2219-T851 and 2519-T87 Aluminum Armor Alloys for Use in Army Vehicle Systems," MTL-TR-87-28, June 1987.

[23] Lincoln Electric, "What You Should Know about Welding Aluminum," *Welding Journal*, Jan. 2000; pp. 54-58.

[24] Polmear, Ian, *Light Alloys From Traditional Alloys to Nanocrystals*, 4th ed., Butterworth-Heinemann/Elsevier, Oxford, UK (2006).

[25] DeVincint, S.M., Devletian, J.H. and S.A. Gideon. "Weld Properties of the Newly Developed 2519-T87 Aluminum Armor Alloy," *Welding J.*, pp 33-43, July 1988.

[26] Nicholas, M., and A. Deschamps, "Characterisation and modelling of precipitate evolution in an Al-Zn-Mg alloy during non-isothermal heat treatments," *Acta Materialia*, 51 (2003), 6077-6094.

[27] Reboul, M.C., B. Duboist, and M. Lashermes, "The Stress Corrosion Susceptibility of Aluminum Alloy 7020 Welded Sheets," *Corrosion Science*, Vol. 25, No. 11, pp. 999-1018 (1985).

[28] Reboul, M.C. and J. Bouvaist. "Exfoliation Corrosion Mechanisms in the 7020 aluminum alloy," *Werkstoffe und Korrosion* 30, 700-712 (1979).

[29] MIL-STD-1946A. Welding of Aluminum Alloy Armor. 1988.

[30] *Damage Tolerant Design Handbook*, Vols. 4 & 5, CINDAS/USAF CRDA Hndbks Operation, West Lafatette, IN, May 1994.

9

Aluminum Alloys: Fabrication, Characterization and Applications II
Edited by: Weimin Yin, Subodh K. Das, and Zhengdong Long
TMS (The Minerals, Metals & Materials Society), 2009

Development of twin-belt cast AA5XXX series aluminum alloy materials for automotive sheet applications

P.Z. Zhao[1], T. Anami[1], I.Okamoto[1], K.Mizushima[1],K.M.Gatenby[2], M.Gallerneault[2], S.Barker[2]
K.Yasunaga[3], A.Goto[3], H.Kazama[3], N.Hayashi[3]
[1]Nippon Light Metal Company, Ltd. NRDC. 1-34-1 Kambara, Shimizu-Ku, Shizuoka-City, Shizuoka-Ken, 421-3203, Japan
[2]Novelis Global Technology Center. Box 8400,945 Princess Street, Kingston, Ontario, K7L5L9, Canada
[3]Honda R&D Co., Ltd. 4630 Shimotakanezawa, Haga-machi, Haga-gun, Tochigi, 321-3393, Japan

Keywords: Twin-belt cast, AA5XXX, Automotive sheet, Microstructure, Stretchability.

ABSTRACT

Process routes for 5XXX series aluminum alloy sheet produced with an advanced twin-belt caster (FLEXCASTER™) have been successfully trialed, and are now in production. The FLEXCAST™ 5XXX sheet has finer intermetallic compounds and grains compared to conventional DC processed 5XXX aluminum alloy sheet, as a consequence of the higher cooling rate during solidification. Optimization of composition and refinement of microstructure has resulted in superior dome stretchability and lower susceptibility to SCC than DC5182 sheets. Moreover, FLEXCAST 5XXX aluminum alloy sheet shows good performance in adhesive bonding and coating tests, critical for automotive structural parts.

INTRODUCTION

In an ongoing effort to reduce the impact on the global environment, automotive companies world wide are making substantial efforts to reduce gas emissions from their vehicles. Reducing the vehicle weight is one of the ways being used to achieve this objective. Aluminum alloys have great potential for replacing steel in automotive applications because of its light weight and ease of recycling [1]. However, formability of aluminum alloy sheets is generally inferior to steel, limiting its application in the automotive industry. Therefore, development of aluminum sheet with excellent formability is needed. In addition, an environmentally friendly fabrication route with reduced energy consumption is required to produce aluminum alloy sheet for mass–produced vehicles. Continuous casting technologies are being developed to meet this need [2,3]. Twin-belt casting is one of the more promising fabrication methods. In particular, the high cooling rate during solidification produces finer intermetallic compounds and grain sizes than with DC cast ingots, resulting in improved formability.

A 1st generation twin-belt casting process was successfully trialed for producing AA5182 aluminum alloy sheets for automotive parts [4]. However, hot rolling was necessary due to the thick slab gauge. Recently, an advanced twin-belt caster (FLEXCASTER™) has been set up by Nippon Light Metal Company to cast thinner gauge slabs without the need for hot rolling [5].

The objective of this study is to develop FLEXCAST™ 5XXX series aluminum alloy sheets for automotive applications. Microstructure, formability, stress corrosion cracking, adhesive bonding and coating performance, critical for automotive structural parts, are evaluated and compared with conventional DC processed materials.

EXPERIMENTAL PROCEDURES
Preparation of materials

The chemical compositions of alloys used in this study are listed in Table 1. The FLEXCAST 5XXX sheet was cast with cooling rates in the range of 20-150°C /s.

Table 1　Compositions of aluminum alloys used in the present study (mass%).

I.D.	Mg	Fe	Si	Mn	Cr	Ti	Al
FLEXCAST 5XXX	3.37	0.20	0.08	0.01	0.01	0.02	Bal.
DC5182	4.65	0.13	0.08	0.21	0.03	0.02	Bal.
DC5052	2.59	0.27	0.09	0.04	0.18	0.02	Bal.

The slab was cold rolled to final gauge (1mm) and then annealed at 460°C on a continuous annealing line. The comparison tests used 1mm thick DC5182 and DC5052 aluminum alloy sheets fabricated via a conventional DC process.

Evaluation of properties

Samples were prepared in accordance with shape No.5 in the JIS Z2201 specification. The 0.2% yield strength (YS), ultimate tensile strength (UTS), total elongation (EL), and r-average value at 15% plastic strain were measured by tensile testing specified by JIS Z2241. The stretching limit dome height (LDH) was measured by dome testing with a 100mm diameter hemispherical punch installed in a stamping machine. The test piece dimension was 200mm square, and the forming strain speed was about 1/s. The load-displacement curve was recorded and the displacement corresponding to maximum load was read as LDH.

Stress corrosion cracking (SCC) was also evaluated. To increase the susceptibility to SCC, 1mm sheets were additionally cold rolled to 0.7mm and sensitized at 120°C for 168 hours. They were then bent into a U-shape and immersed in 3.5%NaCl solution at a current density of 6.2mA/cm^2. The time to failure was taken as the SCC lifetime.

To evaluate the bonding performance, samples were bonded using structural adhesive and exposed to salt spray for 480 hours. The shear strength and morphology of fracture gave a measure of the bonding performance.

To evaluate the coating performance, samples were coated with a paint film and then observed after immersion in hot water for 240 hours.

Evaluation of Microstructures

The grain structure of longitudinal sections was observed by polarizing microscope after buff grinding and anodizing. After argon-ion polishing, a detailed analysis of the crystalline orientation was done by SEM-EBSP. Intermetallic compounds were observed by optical microscope and the size was measured with an image analyzer.

β phases precipitated during sensitization were observed in detail after etching the samples in 1%NaOH solution.

The fractures of stretch forming samples were observed by SEM. Shear bands were checked by etching the formed samples in 10% H$_3$PO4 solution after heating at 120°C for 168 hours.

RESULTS and DISCUSSION
Microstructure

12

Figure 1 shows the microstructures of the FLEXCAST 5XXX, DC5182 and DC5052 samples. FLEXCAST 5XXX has finer grain structure and intermetallic compounds than either DC5182 or DC5052. The measured average grain size was 10μm, 22μm, 19μm, respectively. The distribution of intermetallic compound particles is shown in Figure 2.

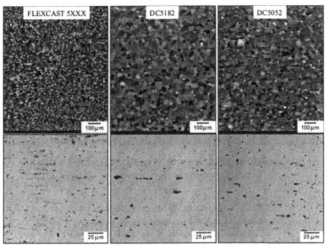

Fig.1   Grain structures and intermetallic compounds in FLEXCAST 5XXX, DC5182 and DC5052 aluminum alloy samples in the longitudinal direction.

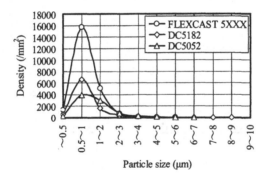

Fig.2   Distribution of intermetallic compounds in FLEXCAST 5XXX, DC5182 and DC5052 aluminum alloy samples.

The FLEXCAST 5XXX sample has a large number of 1μm particles which act as nucleation sites for recrystallization and pinning of the grain boundaries [6], resulting in the fine grain size.
Figure 3 shows the crystalline orientation maps measured by SEM-EBSP. The area fraction of the major orientations are listed up in Table 2. The FLEXCAST 5XXX sample has a higher volume of {112}, {110} and {111} oriented grains than either DC5182 or DC5052, indicating a different mechanism for recrystallization.
The distribution of subgrain and grain boundary misorientation angle is shown in Figure 4. All samples have almost the same distribution of misorientation. 90% of the grain boundaries have a

misorientation angle greater than 15 degrees. Therefore, recrystallization can be considered to be almost completed. The size distribution of grains with more than 15 degree misorientation angle is shown in Figure 5. The grain size of FLEXCAST 5XXX is concentrated around 10μm, while the DC5182 and DC5052 samples have a grain size distribution with a peak of around 30μm and 20μm, respectively.

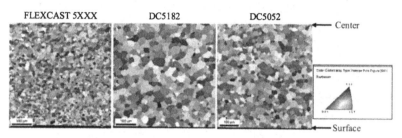

FLEXCAST 5XXX   DC5182   DC5052   ⟵ Center

⟵ Surface

Fig.3   Orientation image maps of FLEXCAST 5XXX, DC5182 and DC5052 aluminum alloy samples in the longitudinal

Table2   Area fraction of major orientation (%).

I.D.	{100}	{112}	{110}	{111}
	11.5	41.2	18.1	14.1
DC5182	12.8	40.4	16.2	11.5
DC5052	16.4	36.2	14.8	8.1

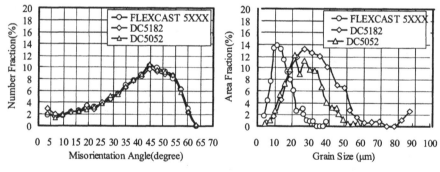

Fig.4   Distribution of grain boundary misorientation angle in FLEXCAST 5XXX, DC5182 and DC5052 samples.

Fig.5   Distribution of grain size in FLEXCAST 5XXX , DC5182 and DC5052 samples.

Mechanical properties and formability

Table 3 summarizes the measured YS, UTS, EL, r-average, and LDH values for the three alloy sets. FLEXCAST 5XXX samples have the highest YS due to the fine grain structure. However, its UTS value was lower than DC5182 due to lower Mg concentration. The r-average value of FLEXCAST 5XXX samples was higher than DC5182 and DC5052 due to the higher volume of {112}, {110}, and

{111} oriented grains.

Table3   Tensile properties and stretching formability of 1mm gauge sheets.

I.D.	YS (MPa)	UTS (MPa)	EL (%)	r-avg	LDH (mm)
FLEXCAST 5XXX	130	235	29	0.80	30.5
DC5182	118	260	29	0.66	29.5
DC5052	101	204	27	0.65	27.4

Surprisingly, the LDH value for FLEXCAST 5XXX was higher than DC5182. In order to understand the difference, the section of broken samples were observed in detail after dome testing. The results are shown in Figure 6. Shear bands in FLEXCAST 5XXX were not as visible due to the lower Mg concentration and finer grain size. Figure 7 shows the morphology of fracture. The FLEXCAST 5XXX sample has smaller dimples than DC5182, indicating that the smaller intermetallic particles suppress void nucleation and growth.

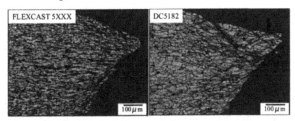

Fig.6   Shear bands observed in FLEXCAST 5XXX and DC5182 samples after dome testing. Shear bands are clearly visible in DC5182.

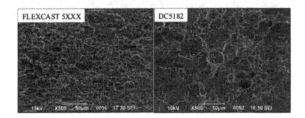

Fig.7   Fracture of FLEXCAST 5XXX and DC5182 samples after dome testing. Dimples in FLEXCAST5XXX are much smaller than in DC5182.

Stress corrosion cracking

Figure 8 shows the appearance of the SCC test pieces for FLEXCAST 5XXX and DC5182 samples after 960 minutes and 10 minutes, respectively. Rupture in FLEXCAST 5XXX samples occurred at the boundary between the aluminum sample and insulation seal, where current density is highest. In contrast, DC5182 samples failed at the top of the loop where stress is considered to be maximum.

15

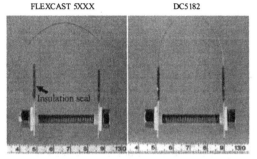

Fig. 8　Appearance of test samples after SCC testing.

Micrographs of the fractured surfaces are shown in Figure 9. DC5182 shows the typical SCC morphology, while FLEXCAST 5XXX shows many etch pits, indicating the sample was broken by metal dissolution rather than by SCC. β phases precipitated during sensitization are shown in Figure 10. It is obvious that the precipitate morphology at the grain boundaries differs between FLEXCAST 5XXX and DC5182 samples. The former precipitates β phases in the form of particles, while the latter precipitates β phases as a film. This is the reason why DC5182 was more susceptible to SCC [7].

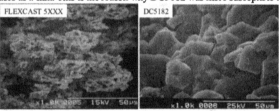

Fig.9　SEM photos of fracture after SCC testing.

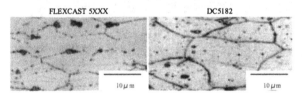

Fig.10　β phases precipitated along grain boundaries after sensitization
C　　h　.

Adhesive bonding and coating performance

Table 4 summarizes the results of adhesive bonding and coating performance tests with FLEXCAST 5XXX and DC5XXX samples containing the same amount of Mg. They have almost the same shear strength and same cohesion failure model after salt solution spray testing, demonstrating that the adhesive bonding performance of FLEXCAST 5XXX is equivalent to that of DC 5XXX. In addition, both materials show good coating performance based on the results of adhesion testing in hot water.

Table4   Evaluation results of adhesive bonding and coating performance.

I.D.	Adhesive bonding	Coating
	Shear strength / Morphrogy of fracrture (salt spray for 480hr)	Adhesion of coating film immersion in hot water for 240hr)
FLEXCAST 5XXX	12MPa/cohesion failure	GOOD
DC5XXX	11MPa/cohesion failure	GOOD

Stamping trials

FLEXCAST 5XXX aluminum sheet has been used successfully to stamp the HONDA hood inner of a ACURA RL, as shown in Figure 11. This was the first trial application of FLEXCAST 5XXX for automobile manufacturing, and the material is now in production. FLEXCAST 5XXX is expected to be used to manufacture other automotive parts in the future.

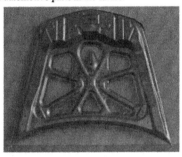

Fig.11   Hood inner of a ACURA RL stamped from FLEXCAST 5XXX aluminum alloy sheet.

CONCLUSIONS

(1) FLEXCAST 5XXX aluminum alloy sheet has finer intermetallic compounds and grains than conventional DC5XXX sheet. This is a consequence of the high cooling rate during solidification.
(2) Optimization of composition and refinement of microstructure of FLEXCAST 5XXX aluminum alloy sheets results in superior dome stretchability and lower susceptibility to SCC than DC5182.
(3) FLEXCAST 5XXX aluminum alloy sheet shows good adhesive bonding and coating performance.

REFERENCES

1. Aluminum International Today, January/February, 2004, 21-30.
2. S.Kim and A.Sachdev:Aluminium Alloys: Fabrication,Characterization and Applications, *TMS*, (2008),51-59.
3. M.Gallerneault and J.D.Lloyd: *EUROMAT2001*,Rimini,Italy,2001.
4. P.Z.Zhao, T.Moriyam, K.Yasunag and N.Hayashi: Proceeding of the 97th Japan Institute of Light Metal, (1997), 209-210.
5. Press release by Nippon Light Metal Company, Ltd., 2004.
6. F.J.Humphreys: Acta Met., 25 (1977), 1323-1344.
7. E.H.Dix, W.A.Anderson and M.Byron Shumaker: Corrosion, 15 (1959), 19-26.

Aluminum Alloys: Fabrication, Characterization and Applications II
Edited by: Weimin Yin, Subodh K. Das, and Zhengdong Long
TMS (The Minerals, Metals & Materials Society), 2009

USE OF ALUMINUM IN ELECTRICAL APPLICATIONS

Peter Pollak, PE
Aluminum Association, Inc.
1525 Wilson Blvd., Suite 600, Arlington, VA 222090

Keywords: Aluminum, Copper, Wire and Cable, Bus Bar, Electrical, Substitution

Abstract

The global commodity boom has increased prices for copper and to a much lesser extent aluminum. Users once again are exhibiting renewed interest in using aluminum for electrical applications in place of copper. The structure of the global economy, mining, electrical equipment, wire and cable industries has changed dramatically, since the last time copper prices reached the price levels we are now seeing.

With the exception of electrical utility transmission and distribution professionals, there is little understanding of aluminum's electrical capabilities with the result that aluminum and its alloys aren't as widely used for electrical applications as they should be. This paper provides technical information about the relevant physical properties and characteristics of copper and aluminum. It describes how aluminum conductors can be design to do a better job than copper for most electrical applications.

To give the reader perspective on reasons why aluminum has traditionally been viewed as only a substitute for copper, the overlapping histories of the electrical and aluminum industries are briefly reviewed in the context of the history of metals. Also addressed are common misconceptions about reliability issues and economic considerations.

INTRODUCTION

After the invention of Hall Heroult process for the electrolytic production of aluminum in 1886, the Pittsburg Reduction Company, Alcoa, Inc.'s predecessor, had to find uses for the metal. One barrier was that existing semi-fabricators of brass and copper did not want to introduce competition to their business. Alcoa responded by going into semi-fabricating of sheet, foil, and wire, and the production of cookware. Similarly to grow aluminum usage in electrical applications, aluminum producers got into the electrical conductor business.

Over the years Kaiser, Alcoa, Reynolds, etc. all got out of the wire and cable business and Southwire, General Cable, etc. sold their stakes in aluminum smelters. Today Rio Tinto's Alcan Cable Company is the only remaining wire and cable producer that is part of an aluminum company and RTA has been announced publicly that the Alcan Cable Company is to be sold.

Copper because of its relative ease to produce from ore has been used by mankind for thousands of years. Over time production developed so that it was available in sufficient quantity and cost for a variety of applications. Copper was a basic engineering material thousands of years before the industrial revolution, and the dawn of the electrical industry in 1882.

The history of aluminum is much more recent. Because of its affinity to react and combine with other elements, only small amounts of metallic aluminum were produced during the19[th] century by chemical reduction processes. This made aluminum a rare, precious metal, more valuable than gold, silver and platinum.

The history of electricity and aluminum are intertwined, because electricity is needed to electrolytically produce aluminum. Thus aluminum was not available to build electrical equipment at the beginning of the electrical industry. Readily available copper was used, because it provided the best, cost-effective conductivity of available materials at the time.

The current situation is vastly different: today world production of aluminum is roughly double that of copper; and among metals global production of aluminum is only second to steel. The price per pound of metals fluctuates with supply and demand. Over the past few years copper prices have increased much more rapidly than aluminum prices. As a result, some traditional applications that were once copper today use aluminum. However, the physical characteristics and economics of these metals suggest that there could be a lot more aluminum substitution for copper in electrical applications.

PHYSICAL PROPERTIES OF ALUMINUM AND COPPER

The following chart summarizes the relevant physical properties of aluminum and copper.

Table 1. Physical properties of aluminum and copper

Property	ALUMINUM 1350-H16 [1]	ALUMINUM 6101-T6 [1]	ANNEALED COPPER [1]
DENSITY (Specific Gravity) (lbs/ in**3)	2.70 0.975	2.70 0.975	8.93 0.3225
VOLUME CONDUCTIVITY (%IACS)	61	55	100
WEIGHT CONDUCTIVITY	100%	93%	50%
STRENGTH UTS (MPa) (KSI) YTS (MPa) (KSI)	125 18 110 16	220 32 195 28	235 34 104 15

From an engineering perspective, the most important properties and characteristics for conductor material selection are: density, electrical conductivity, strength and corrosion resistance.

Density

Aluminum has roughly one third the density of copper. For the same weight, aluminum provides three times the volume of copper.

Electrical Properties

Annealed copper has a conductivity of 100% IACS and 1350-H116 has a conductivity of 61% IACS; so that, for a given cross-section, copper provides greater conductivity. However, equivalent conductivity can be achieved with a larger cross-section of aluminum. Thus aluminum conductors are always larger than their copper electrical equivalents. Generally two AWG sizes larger in the smaller wire sizes and 50% larger in cross-sectional area for the larger conductors sizes. Because aluminum, however, has a much lower density than copper, the larger aluminum conductor weights much less than its copper equivalent. The result is that one pound of aluminum can provide the same conductivity as 2 pounds of copper. (This is based upon aluminum alloys having 61% IACS conductivity; slight adjustment is needed if the conductivity is less.) When space permits, this is the basic reason why aluminum is used in place of copper for current carry applications, such as wire and cable, transformer strip and bus bar.

Strength

Generally speaking copper is stronger than aluminum for a given cross section. However, the strength of aluminum can be increased by alloying and processing. For pulling wire and cable into conduit only half the weight of the equivalent copper is involved. To provide the mechanical strength required by bus conductors during short circuits, higher strength alloys with slightly lower electrical conductivity in the 6xxx series are used. For transmission cable where long-distance spans between towers require more strength, a composite, aluminum-conductor-steel-reinforced (ACSR) cable, is used.

Corrosion Resistance

When aluminum is exposed to air the oxide coating created seals aluminum from further contact with the air and prevents continuing oxidation. If aluminum is scratched, it self-seals at that location when nascent aluminum is exposed to the air, preventing further corrosion. The oxide coating is insulating; however, it is very thin and brittle. The coating can easily be disrupted to make a good electrical connection.

If the surface oxidation layer is disturbed in a moist environment that does not allow for direct contact with the air, corrosion will occur. Photo 1 illustrates a failure that occurred in a direct buried cable when it was unknowingly nicked during installation. The cable failed after approximately four years of service. The failure shown in Photo 1 was taken approximately five years after the failure. It is noteworthy that despite the aggravated state of corrosion (white powder) in the failure, once the adjoining conductor was cut to remove the faulted section, surface oxidation formed quickly on the areas of the cut conductor exposed to the air, and no further corrosion occurred.

Photo 1

Listed below is empirical data illustrating the atmospheric corrosion rate in average Mils per year for solid metals that are used in the electrical industry.

Table 2. Atmospheric corrosion of various solid metals used in electrical construction [1]

ATMOSPHERIC CORROSION OF SOLID METALS OVER A 10 YEAR PERIOD					
Atmosphere	Location	Al	Cu	Zn	Pb
Desert	Phoenix, AZ	0	0.005	0.01	0.009
Rural	State College, PA	0.001	0.023	0.042	0.019
Coastal	Key West, FL	0.004	0.02	0.021	0.022
Coastal	La Jolla, CA	0.028	0.052	0.068	0.016
Industrial	New York, NY	0.031	0.047	0.19	0.017
Industrial	Altoona, PA	0.025	0.046	0.19	0.027

MISCONCEPTIONS

There are misconceptions regarding aluminum's suitability for use in electrical applications. While the physical properties of aluminum are well known, other issues have lead to misunderstanding and misconceptions.

Reliability

Aluminum conductors are reliable. All transmission and distributions conductors, and most service entrance and large feeder cables are aluminum. They have a long history of reliable service.

However, all types of aluminum wire suffer from association with residential branch circuit wiring produced in the 1960s and 1970s. Solid aluminum wire sizes #12 and #10 AWG were used for branch circuits in hundred of thousands of homes during the late 1960s and early 1970s. The Consumer Product Safety Commission (CPSC) [3] reacted to reports of overheating connections by conducting an investigation. Their research concluded that aluminum branch-circuit wiring connections were hazardous.

Unbeknown to most people is that the Canadian Government [4] conducted a similar investigation, which concluded that properly installed, aluminum branch circuit wire connections were not hazardous. Furthermore, proposals to change the National Electrical Code (NEC) [5] to eliminate aluminum-branch-circuit wiring were rejected by the electrical experts on Code Making Panels and by Underwriters Laboratories (UL) [6].

One can debate if the CPSC or Canadian Board of Inquiry conclusion was correct. However, the performance history of the aluminum wiring in question over the past three-to-four decades speaks for itself. It is still installed in hundreds of thousands of US homes today and continues to perform its intended function. It appears that the Canadian Board of Inquiry, the National Fire Protection Association NEC Code Panel members, and Underwriters Laboratories were correct in their determination which permitted continued use. Even though the National Electrical Code still permits aluminum branch circuit wiring, it hasn't been produced in decades because of the bad publicity.

History

Aluminum has only been available in sufficient quantities for expanding electrical usage for the last 60 years. (Ninety five percent of all the aluminum ever produced was made since WWII.) During the 60 years prior to that time, the electrical industry evolved using copper. This is still evident from the benchmark used to measure electrical conductivity for all materials – "% IACS". The International Annealed Copper Standard (IACS) was adopted by the International Electro-technical Commission (IEC) [7] in 1913 using the volume conductivity of annealed copper for the 100% base.

Because copper was used to the build the original US electrical infrastructure during the 19th and most of the 20th century, it became the de facto standards for electrical conductors. Aluminum coming later had to be a substitute for copper in conductor applications. Where aluminum is used in place of copper, it can be undeservedly perceived as just a "cheap substitute".

Economics

Engineers like most people know there is no "free lunch". With few exceptions, you get what you pay for. If something appears too good to be true, it usually is. Because of this, aluminum is often perceived as not as good as more expensive copper for which it can be substituted. Since history and the new economics of base metals aren't widely known, many would argue that aluminum must be inferior, since it costs less. The fact is that material prices vary with global supply and demand. However, material capabilities are what they are, and do not change with supply and/or demand.

23

Substitution Psychology

The psychology that accompanies substitution can also mislead. For example when a problem develops, search for a cause to prevent recurrence is worthwhile. However all too often, anything that appears different can be mistakenly taken as the cause of the problem. If aluminum substitution has occurred and a problem develops, it is often incorrectly attributed as the cause without thorough investigation and analysis, which may show that the problem is unrelated to aluminum and would have occurred had copper been used. The rush-to-judgment phenomena can be bolstered by the "cheap" and "substitute" perceptions.

SUMMARY

Both the aluminum and electrical industry are about 120 years old; however, during the first 60 years relatively little aluminum was produced. It is only since WWII that aluminum production dramatically increased. More than 95% of the aluminum ever made was produced after WWII. The electrical industry initially used copper because it is a good conducting material that was readily available.

Thus copper became the de facto standard for electrical conductors. Since WWII aluminum has replaced copper in many electrical applications. Today roughly twice as much aluminum as copper is being produced.

CONCLUSION

Aluminum is now available for electrical applications at costs much lower than for copper. If the electrical industry were starting today, aluminum would be the standard material for conductors because of its physical attributes, availability and the economics resulting from global industrialization. Copper will continue to be used for electrical applications because of historical momentum and misperceptions. Aluminum will be increasingly used in place of copper for existing and new electrical conductor applications.

REFERENCES
[1] Aluminum Electrical Conductor Handbook, Aluminum Association, Arlington, VA
[2] Electrical Wire Handbook, The Wire Association International, Guilford, CT
[3] US Consumer Product Safety Commission 4330 East West Highway, Bethesda, MD 20814
[4] Ontario Government Inquiry into Aluminum Wiring Dr. J. Tuzo Wilson, Commission Chair, Queen's Printer for Ontario
[5] National Electrical Code, currently published by the National Fire Protection Association, Batterymarch Park, Quincy, MA 02169-7471
[6] Underwriters Laboratories, Inc., 333 Pfingsten Road Northbrook, IL 60062-2096
[7] International Electro technical Commission, IEC Central Office, 3 rue de Varembe, P.O. Box 131, CH – 1211 Geneva 20, Switzerland

Aluminum Alloys: Fabrication, Characterization and Applications II
Edited by: Weimin Yin, Subodh K. Das, and Zhengdong Long
TMS (The Minerals, Metals & Materials Society), 2009

Re-Use of Aluminum Turning Chips by Hot Extrusion

Dirk Biermann[1], Klaus Pantke[1], A. Erman Tekkaya[2], Marco Schikorra[2],

[1] Technische Universität Dortmund, ISF, Department of Machining Technology, Baroper
Str. 301; 44227 Dortmund, Germany

[2] Technische Universität Dortmund, IUL, Institute of Forming Technology and Lightweight
Construction, Baroper Str. 301; 44227 Dortmund, Germany

Keywords: Extrusion, Aluminum, Recycling, Chips

Abstract

Aluminum is the most widely used metal after steel. Due to its convenient material properties, this material can be used in several products. The high energy requirement for mining
and melting aluminum is one of the major disadvantage of this material. Although, the
re-melting of aluminum scrap can reduce the energy requirements, the needed energy for the
melting process itself is still high. This article presents a process chain of a direct conversion
technology of aluminum chips by cutting, compaction of the chips to billets, hot extrusion to
a rectangular square profile, and finally characterization of the profile properties. It is shown
that by the direct conversion, melting the chips for secondary use may become unnecessary.
Due to the fact that this process chain doesn't need a melting process, there will be great
advantage over the conventionally process chain to save ecological and economical resources.

Introduction

A reduction in energy needed for production processes is becoming of increasing importance
in order to prevent continuously rising energy costs as well as, developments like global
warming caused by greenhouse effect [1]. Especially, energy intensive production processes
like molding are eminent topics for energy saving. Aluminum is one of the most important
light metals. Therefore aluminum is used frequently in aviation industry. For safety reasons,
parts for aviation constructions are made often completely by a machining process. The
resulting chips are normally supplied as a raw material to a melting process for recycling.
The remelting of aluminum scrap is already a energy-saving way for aluminum recycling
in contrast to extract aluminum by electrolysis of bauxite. Nevertheless, the remelting of
aluminum scrap requires significant energy. For example, the electrolytic extraction from
bauxite needs about 13 to 15 kWh per kg [2]. This value improves when using scraped
aluminium from recycling. But there still is a loss of up to 20% of the material during the
remelting stage by oxidation on the metal surface, burning and mixture with the slag. In
order to avoid this energy intensive process, a new process chain for aluminum scrap was
developed. This process combining secondary material usage and a reduction of process
steps, is the direct conversion technique dealing with a reuse of scrap-chips from milling or
turning processes and hot profile extrusion. This paper gives details of the process chain by
describing turning and milling chips, compacting of these chips and finally their hot extruded
profiles. Figure 1 shows an overview of the process chain.

As presented in figure 1, the first component of this process chain is the classification
of the resulting chips from milling and turning processes. In the next step, these chips are
compressed and hence compacted to billets and finally, they are hot extruded to new profiles.
After this, the aluminum profiles can be used for further applications. The recycling chain
can be closed, if these profiles are machined using cutting tools again.

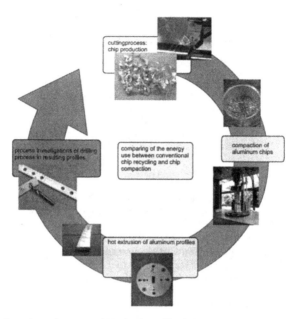

Figure 1: Overview of process chain with reuse of aluminum scrap by hot extrusion

Characterization and production of Aluminum chips

For the investigation and the prevention of inadequate chip forms in the presented research, two different cutting processes were selected. In industrial production, turning and milling are the most commonly used machining operations. Therefore, an efficient recycling of resulting chips can present a promising alternative to the conventional melting process. The resulting waste product from milling and drilling processes contains chips of different geometric forms. This chip-form is an important factor for the supplementary compaction process. Several problems in the compacting step are encountered due to inadequate geometric chip forms. Chips with snarled, long and helical geometries can stick together quite easily. In contrast to that, compacting elemental or needle chips can be more difficult [3]. To handle these problems, chip forms in connection to different machining options and setting parameters were investigated. The cohesion of needle chips is critical and the following extrusion process is complicated.

For example, turning of aluminum-alloys principally generates long chips. This is due to the continuous cut in the turning process. Due to this, the cutting edge during the process is continuously in contact with the material resulting in a material removal that influences the chip-form. If the resulting chip doesn't break, long chip-geometries accrue. To manipulate the resulting chip-form, process parameters have to be adjusted. In contrast to turning, the milling process is a discontinuous cutting operation. During the process, the cutting edge is one time in contact with the material and generates a chip and right after that, the same edge is no more in contact with the material and a chip breaking will automatically be occurred. Therefore, milling of aluminum produces elemental or needle chips as compared to the chips formed during turning. For the further investigations on compacting and hot extrusion only

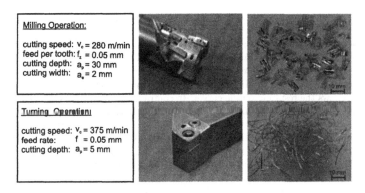

Milling Operation:

cutting speed: v_c = 280 m/min
feed per tooth: f_z = 0.05 mm
cutting depth: a_p = 30 mm
cutting width: a_e = 2 mm

Turning Operation:

cutting speed: v_c = 375 m/min
feed rate: f = 0.05 mm
cutting depth: a_p = 5 mm

Figure 2: Example of chip forms depending on different machining processes

milling chips were used. For this work, chips of the aluminum alloy AA6060, AA6062 and AA7075 were used.

Compaction to Billets

The resulting chips are compacted using a conventional universal testing machine. In the compaction process, the chips are charged into a metallic cylinder. This cylinder is clamped on the testing machine, before is compacting process. Each compaction step was stopped when a maximum force of 60 kN was reached. After removing the punch, the resulting free volume inside the tube was filled with additional chips and compaction was repeated until a billet length of approximately 145mm ± 5mm was reached. The number of compacting steps depends on the alloys used as well as on the resulting chip geometry. In the compaction step, billets with the alloy AA6060 as well as the mixed one with AA6060/6082 and AA6060/7075 were prepared.

Hot Extrusion of Aluminum Profiles

For the extrusion of the compacted billets, a 2.5 MN extrusion press was used. The pre-compacted billets as well as conventionally cast billets were preheated before extrusion to an initial temperature of 500C. As process conditions, a container temperature of 450C and a constant ram speed of 1 mm/s were, used.

Figure 3: Hot Extrusion of billets [3]

27

Results and Conclusion

The presented investigations can show that a hot extrusion of aluminium scrap after a compaction step without a remelting process is possible. To evaluate the mechanical properties of these profiles in contrast to casted and extruded profiles, tensile test specimens were prepared from the extruded profiles and test were carried out under quasi-static conditions. Figure 4 shows the resulting tensile curves. In contrast to the conventional extruded profile, the yielding of a profile form AA6060 chips is marginally lower. The strength of profiles can be increased by mixing the AA6060 chips with faster alloys in the compaction step. Figure 4 shows that the strength of mixed chips from AA6060 and AA6082 as well as AA6060 and AA7075 was increased significantly. Profiles with mixed chips AA6060/7075 can have an increased strength up to a value of 280 MPA at 15% true strain.

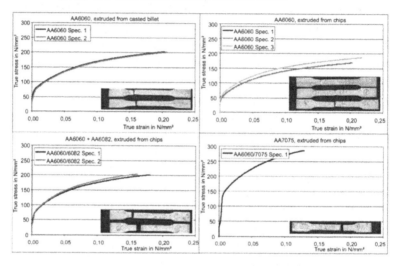

Figure 4: Tensile test data and specimen for aluminium chip mixtures [4]

References

[1] Baumert,K., Herzog, T., Pershing, J.: Navigating the numbers- greenhouse gas data and international climate policy. Report of the world resources institute, 2005.
[2] John, A., Green, ed.: Aluminum recycling and processing for energy conservation and sustainability. ASM International, Ohio USA, 2007.
[3] Tekkaya, A.E., Biermann, D., et al.: Hot Profile Extrusion of AA-6060 Aluminum Chips, Journal of Materials Processing Technology, Elsevier, 2008.
[4] Schikorra, M., Pantke, K., et al.: Re-use of AA6060, AA6082 and AA7075 aluminum turning chips by hot extrusion, Proceedings of ICTP conference 2008, Gyeongju Korea, 2008.

Aluminum Alloys:
Fabrication, Characterization and Applications II

Processing and Properties

Session Chair

Gyan Jha

Aluminum Alloys: Fabrication, Characterization and Applications II
Edited by: Weimin Yin, Subodh K. Das, and Zhengdong Long
TMS (The Minerals, Metals & Materials Society), 2009

Multiple extrusion and consolidation of Al-4Mg-1Zr

Daniel Aguilar Garcia[1], Richard J. Dashwood[2], Martin Jackson[3], David Dye[1]

[1]Department of Materials, Imperial College London, Prince Consort Road, London SW7 2BP, UK
[2]Warwick Manufacturing Group, International Manufacturing Centre, University of Warwick, Coventry CV4 7AL, UK
[3]Department of Engineering Materials, The University of Sheffield, Sheffield S1 3JD, UK

Keywords: Aluminium, Al-Mg-Zr, Severe Plastic Deformation, Co-extrusion.

Abstract

In the continued quest for metallic alloys with better properties, metallurgists have employed a variety of thermo-mechanical processing routes and non-conventional methods. While steady progress has been made in this area, recent work promises to produce alloys with a step change in properties via severe plastic deformation (SPD) techniques. Several SPD techniques are now being studied such as Equal-Channel Angular Pressing (ECAP), High Pressure Torsion (HPT) and Accumulative Roll Bonding (ARB). In this paper a new SPD technique is applied to a novel experimental alloy, Al-4Mg-1Zr. The alloy has been subjected to six passes of conventional extrusion. The mechanical properties and the microstructure was studied after each pass. Analysis showed that after the each pass the microstructure has been refined, the primary Al_3Zr particles were broken down and the hardness increased slightly. However, the yield stress and the ultimate strength increased significantly after the first pass and decreased for the following passes.

Introduction

Aluminium alloys are widely used in the aerospace industry as they provide a good combination of specific strength and low cost. The high strength and enhanced superplastic properties associated with ultra-fine microstructures in many alloys provide substantial benefits for a wide range of applications. Obtaining ultra-high strength materials will enable component sections to be reduced, leading to weight savings, reduced fuel consumption and lower CO_2 emissions. The combination of properties is important, but it has always been a challenge to produce very high strength alloys using commercial processes. A number of processing routes have been explored to obtain amorphous and nanocrystalline alloys. Both the primary synthesis process by which the initial structure is obtained as well as the secondary processing route to obtain a bulk shape is important. Primary processes have included powder atomisation, mechanical alloying, or inert gas condensation. All of these are rapid solidification processes, which are required to obtain the desired structures. Coupled with this is alloy development. Secondary processing routes such as conventional extrusion, equal channel extrusion or torsion straining have been widely reported (1; 2; 3). The challenge with secondary processing routes is to maintain the amorphous or nanocrystalline structure, because as temperature and deformation are applied, the microstructure tends to recrystallize and coarsen (4; 5). An alternative processing route that has been demonstrated at the laboratory scale is severe plastic deformation, in which a conventionally cast alloy is deformed through processes such as equal channel angular extrusion or pack rolling to produce a highly refined structure (6). The decision to exploit a multi-extrusion method is due to the extremely good properties that might be potentially generated. A material that has been subjected to severe plastic deformation would be obtained, thus improving mechanical properties, such as hardness and yield strength (7). Extrusion leads to a substantial change in the strength of the material: both hardness and tensile yield strength increase (7). In conventional extrusion, the degree of work and thus grain refinement are quite limited, hence grains and inclusions cannot be refined to a great extent by this process. This report describes how a multi-extrusion technique was employed to produce ultra-fine grained material. Multiple extrusion passes should generate a very large amount of plastic strain that can be accumulated in the sample without substantially changing its dimensions after the first pass. The experimental alloy Al-4Mg-1Zr, was extruded six times, refining its microstructure and improving its mechanical properties during each pass.

Experimental

A new experimental alloy, Al-4Mg-1Zr (8), was chosen for the multi-pass extrusion (Fig. 1) studies. The received aluminium needles were cold compacted into a solid extrusion billet, 120 mm high and 72 mm in diameter. For the first pass the billet received a pre-heat of 30 minutes at 500°C and was extruded using a ratio of 20:1 at a speed of 1 mms^{-1}, the rest of extrusion passes were carried on at 450°C. After every extrusion pass the specimen was cut into 120 mm long rods that were packed together in a steel can for pre-heating before the next extrusion. The extrusions presented in this study were all performed on a 5 MN Enefco hydraulic press at Imperial College London.

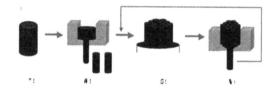

Figure 1: Multiple extrusion pass process, a) Al-4Mg-1Zr extrusion billet of cold compacted needles b) Hot extrusion of the billet c) Rods cut from the extrusion and d) Next extrusion pass.

For microstructural analysis all the samples were prepared by cutting, grinding, polishing and anodizing. They were micro-anodized at 20 V in Barkers solution (2% tetraluoroboric acid, 98% H$_2$O), to reveal the grain structure under cross polarized light. Secondary electron images (SEI) and backscattered electron images (BSEI) were obtained using a Jeol 840A scanning electron microscope (SEM). For transmission electron microscopy (TEM) studies, thin slices were cut from the extrusion rods transverse to the extrusion direction. Thin foils were prepared by electropolishing at -35°C using a solution of 5% perchloric acid and 5% nitric acid in methanol. Micrographs were obtained using a Jeol TEM 2000FX operating at 200 kV.

After each extrusion pass Vickers hardness measurements were made with a 5 kg load on both longitudinal and transverse directions to the extrusion with an Indentec Vickers testing machine model 6030LKV. Hardness results are from an average of 10 measurements.

Tensile testing was carried out on samples with a 6.5 mm diameter and a 24.5 mm gauge length machined from the extrusion in the longitudinal direction. However, the sample gauge length was too small for an extensometer to be used. The strain rate was maintained constant at 5×10^{-4} s^{-1}.

ted at 360°C for 100 hours and tested at 450°C with initial strain rates of 10^{-4} s^{-1}.

Results and discussion

Microstructure evolution

The Al-4Mg-1Zr needles were first cold compacted to a billet for extrusion at 500°C. The microstructure after this first pass consists of a very well defined array of elongated grains with subgrains or recrystalised grains in the extrusion direction, while the face normal to extrusion consisted of relative fine equiaxed grains with some zones of coarse equiaxed grains (Fig. 2). When extruding a metallic material stored energy is imparted due to the hot work and the dislocations density increases. At elevated temperatures part of the stored energy is expected to recover. Dynamic recrystallisation has been observed at elevated temperatures where dynamic recovery takes place (9). It is clear from previous work that recrystallisation in Al-4Mg-1Zr is stimulated by the coarse primary Al$_3$Zr particles (10). The interest is to have a fine dispersion of grain growth restricting metastable cubic Al$_3$Zr particles that cannot be provided by simple extrusion (11). By extruding the material over several passes, the grain size reduces in both the extrusion and traverse direction, as shown in Fig. 2. The microstructure in the extrusion direction after the last pass consists of very thin and elongated grains with a large fraction of sub-micron equiaxed grains (Fig. 3) with an average grain size of 0.5 μm. Figure 6 shows the evolution of the microstructure over the extrusion passes. It can be observed that the equiaxed recrystallised grains appear firstly inside the elongated grains that eventually disappear. The observed microstructural evolution agrees with that described by Gholinia et al. (12)-when a material is deformed in plane strain compression, the microstructure is elongated and the grain boundaries are pushed towards each other. At elevated temperatures where the boundaries are mobile, and at a sufficiently large strain, adjacent grain boundaries may eventually pinch off leaving an approximately equiaxed microstructure. This continous recrystallisation process is often named geometric dynamic recrystalisation (GDR). Aluminium alloys showing a dispersion of fine second phase elements, as in this study, are supposed to inhibit the grain boundary migration necessary for continous recrystallisation raising the critical strain or temperature. It is believed that in this study the critical strain was achieved.

Through the different passes the Al$_3$Zr particles were analyzed in order to determine whether the multiple pass extrusion helped to break the coarse particles into a fine distribution of second phase particles. Figure 4 shows that the coarse particles have been reduced after each pass and more homogenously distributed through the matrix.

Mechanical properties

The mechanical properties were analyzed via tensile and hardness tests (data is shown in Table 1). After the first pass, the yield stress, the elongation to failure and the stiffness of the material increases significantly. In the third pass the yield stress decreased, the stiffness remained constant and the elongation to failure increased. There was no further variation in mechanical properties after subsequent passes. Specimens with saturated solid solutions, when tested at certain temperatures and strain rates, showed a serrated stress-

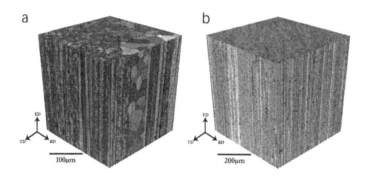

Figure 2: Light micrograph of the a) first extrusion pass and b) sixth extrusion pass.

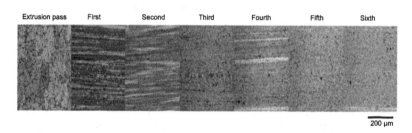

Figure 3: Light micrographs of the microstructure evolution in the extrusion direction.

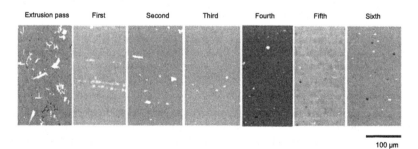

Figure 4: SEM showing images of the primary Al_3Zr (bright phase) in the different extrusion passes.

34

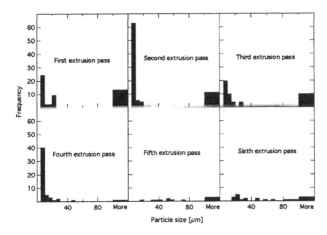

Figure 5: Particle size distribution of Al$_3$Zr in the different extrusion pass.

Figure 6: TEM image of the sixth extrusion pass on the Al-4Mg-1Zr showing (a) fine grains (bright field image), (b) distribution of Al$_3$Zr precipitates (dark field image) and (c) diffraction pattern.

strain curve, associated with the repeated initiation and propagation of deformation bands. This phenomenon is known as the Portevin-Le Chatelier (PLC) effect and is observed as a consequence of dynamic strain aging. This has been associated with the diffusion of solute atoms to arrested dislocations, giving rise to an additive contribution to the flow stress (13). The solute atoms are able to diffuse in the specimens at a rate faster than the speed of the dislocations, so as to catch and lock them. The load increases when the dislocations are torn away from the solute atoms, resulting a load drop until the next band of dislocations interact. The PLC effect reduces the formability of the material because it reduces the strain to necking (14). The hardness increased significantly in the second pass for both directions. In the third pass the hardness decreased in the direction perpendicular to extrusion but slightly increased in the extrusion direction. The results of the first extrusion pass are in agreement with previous work (15).

35

Table 1: Results of the tensile and hardness tests of the extrusion passes

Extrusion Pass No.	Yield Stress MPa	Ultimate Tensile Stress, MPa	Strain to Failure	Hardness Longitudinal, H_v	Hardness Transverse, H_v
1	222	346	18	86	83
2	283	349	21	94	103
3	283	368	24	96	96
4	272	343	27	94	98
5	256	343	25	95	96
6	255	343	24	95	95

Conclusion

The feasibility of imparting severe plastic deformation to a material via a commercial process, extrusion, has been demonstrated. This multiple extrusion pass process not only reduces the microstructure of the material to the micron scale, but also improves the mechanical properties, yield stress and the elongation to failure, maintaining the hardness.

References

[1] Roven, H.J. et al. *Mater. Sci. Eng. A*, vol. 410-411, 2005/11/25.

[2] Furukawa, M. et al. *Mater. Sci. Eng. A*, vol. 410-411, 2005/11/25.

[3] Kaibyshev, R. et al. *Mater. Sci. Eng. A*, vol. 396, 2005/04/15.

[4] Z. Peng and T. Sheppard *Model. Simul. Mater. Sci. Eng.*, vol. 12, 2004/01/.

[5] M. Kassner and S. Barrabes *Mater. Sci. Eng. A*, vol. 410-411, 2005/11/25.

[6] H. Miyamoto, K. Ota, and T. Mimaki *Scr. Mater.*, vol. 54, 2006.

[7] Nagasekhar, A.V. and et al. *Mater. Sci. Eng. A*, vol. 423, 2006/05/15.

[8] R. Grimes, R. Dashwood, A. Harrison, and H. Flower *Mater. Sci. Technol.*, vol. 16, 2000/11/.

[9] M. Kassner and S. Barrabes *Mater. Sci. Eng. A*, vol. 410-411, 2005.

[10] S. Katsas, R. Dashwood, R. Grimes, M. Jackson, G. Todd, and H. Henein *Mater. Sci. Eng. A*, vol. 444, 2007.

[11] Dashwood, R.J. et al. in *Mater. Sci. Forum*, vol. 357-359, 2001.

[12] A. Gholinia, F. Humphreys, and P. Prangnell *Acta Mater.*, vol. 50, 2002.

[13] Klose, F.B. et al. *Mater. Sci. Eng.*, vol. A369 76–81, 2004.

[14] Benallal, A. et al. *Int. J. Plast.*, vol. 24, 2008.

[15] S. Katsas PhD thesis, Imperial College London, 2006.

Aluminum Alloys: Fabrication, Characterization and Applications II
Edited by: Weimin Yin, Subodh K. Das, and Zhengdong Long
TMS (The Minerals, Metals & Materials Society), 2009

AL-ZN-MG FOR EXTRUSION – HOT WORKABILITY

H.J. McQueen[1], Ani Shen[1], P. Leo[2] and E. Cerri[2]

[1] Mech., Indu. Eng., Concordia University, Montreal, H3G 1M8, Canada
[2] Innovation Eng. Dept., University of Salento, 73100, Lecce, Italy

Keywords: hot working, Al-Zn-Mg alloys, constitutive analysis

Abstract

Al-Zn-Mg alloys have lower strength but less quench sensitivity than 7000 series with Cu; they are useful for many terrestrial structural and transport applications, especially since they have good extrudability. As-cast specimens of Al-5.5Zn-1.2Mg were torsion tested to fracture in the ranges 250 - 500°C and 10^{-2} to 5 s^{-1}. The peak strength fell markedly from 300 to 400°C indicative of precipitate formation and coalescence and then slowly up to 500°C due to more solution at higher temperature; it was lower for lower strain rates. The ductility rose rapidly from 300 to 400°C and with decreasing strain rate, but at 500°C, it was high but inconsistent. The constitutive analysis by the sinh equation gave Q_{HW} = 161 kJ/mol and n = 1.05. Optical microscopy exhibited elongated grains with larger subgrains at higher temperature and lower strain rate. In hot tensile tests, flow curve shapes are similar, peak stresses are fairly consistent and ductility (although fivefold less) vary similarly with temperature and strain rate. The tensile activation energy was 218 kJ/mol and n = 1.75. The hot working behavior is compared to those of 7004 and 7020 alloys and of high-strength 7075 and 7012 alloys.

Introduction

The present alloy Al-5.5Zn-1.2Mg (55.12) is a lean precipitation-strengthened alloy with medium strength and ductility at 20°C. Such alloys differ from the 7000 series aircraft alloys in having no Cu which alters the important η phase ($MgZn_2$ towards MgCuAl) that provides very high strength [1]. These Al-Zn-Mg alloys, strengthened by $MgZn_2$ and T phase ($Al_2Zn_2Mg_3$) with low quench sensitivity and solution temperature of about 350°C, can be press quenched by air jets at the extrusion exit [2]. It has higher extrudability than 6061 and when properly aged has higher strength and good corrosion resistance [1-4]. Extrusion studies of 7005 (4.8Zn-1.2Mg-0.3Mn-0.2Zr) [3], 7003 (6.0Zn-0.8Mg- 0.16Mn-0.16Zr) [4] and of high purity, Al-7.5Zn and Al-7.75Zn-2.47Mg [5-6], showed that the microstructures were fibrous; for the last two high purity alloys, the subgrain sizes increase with rising temperature T and falling strain rate $\acute{\epsilon}$ as for Al [5-7]. Its good weldability without further heat treating makes it an alternative to Al alloys with about 5%Mg and 0.7Mn. In hot working [8-13], its flow stresses are expected to be much lower and ductility much higher than the values for 5083 and 5182 [14-17]; comparison is made with these and other Al-Mg alloys [17-19]. Generally, dynamic recovery (DRV) is the sole restoration mechanism in Al alloys [17-20] but it is much attenuated by Mg solute drag even to the extent of permitting particle-stimulated nucleation (PSN) of dynamic recrystallization (DRX) [14-17,20].

Specimens from as-cast billets [8,11] were tested to fracture over the ranges 200-500°C and 0.01 to 5 s^{-1} in torsion, as described elsewhere [14-16,19]. The torque and surface strain were transformed into equivalent stress σ and strain ε by the traditional means. Comparison is made during presentation of the results to the alloy 7004 (Al-3.82Zn-1.12Mg-0.46Mn) [11] that has hot workability previously found very similar to 7020 (4/5Zn-1/1.4Mg-0.1/0.5Mn 0.1/0.35Cr) [9].

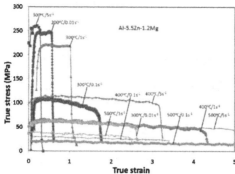

Figure 1. Torsion σ-ε curves for Al-5.5Zn-1.2Mg with high peaks at 200-300°C but broad peaks, high ductility at/above 400°C; stresses lower at lower έ for each T.

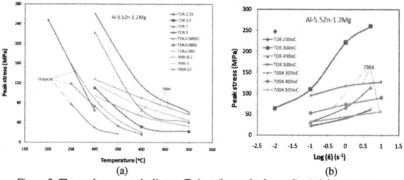

(a)   (b)

Figure 2. The peak stresses decline as T rises (lower for lower έ), a) doing so more gradually above 400°C and becoming more similar to 7004 [9]. The tensile stresses follow a similar pattern but are markedly high for their low έ [8] (final digits give έ as s⁻¹). Peak stress dependence σ_P on έ is almost logarithmic (b) as for 7004 [9].

Results

At 300°C (Figure 1) each σ-ε curve exhibits rapid hardening to a peak and a sharp decline to fracture at a low strain whereas at 400 and 500°C, the peak is followed by a gradual decline to a steady state with a high fracture strain. At each T, the flow curves change in a similar pattern as έ decreases. Flow stress rises and ductility ε_F declines as έ rises. Such shape and variation with T and έ were observed in hot tensile tests on the alloy; however, the ductilities were less by a factor of 5. The flow curves of rolled-plate 7004 exhibit low strains to sharp peaks and then more gradual declines than the present alloy [9].

In Figure 2a, σ_p decreases at a declining rate as T rises and the 0.01, 0.1 and 5 s⁻¹ curves are roughly parallel to 1 s⁻¹ but 5s⁻¹ is closer to and above it. The tensile (TEN) results are also roughly parallel to the torsion (TOR) but they are markedly high with 0.001 s⁻¹ (TEN) similar to

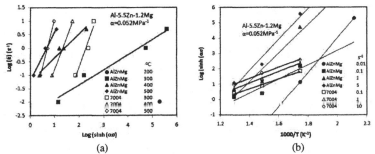

(a)　　　　　　　　　　(b)

Figure 3. In accordance with Eqn. 1, log $\acute{\varepsilon}$ vs. log (sinh $\alpha\sigma$) (a) provides straight lines at each T; however, they are not as parallel (n_{av}= 1.05) as those of 7004 (n_{av}= 2.79). In the Arrhenius plot (b) lines for present alloy are fairly parallel except for 0.1 s^{-1} and have (S_{av}= 8.03) compared to 7004 (S_{av}= 3.47).

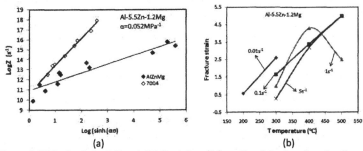

(a)　　　　　　　　　　(b)

Figure 4. With the derived Q_{HW} (=2.3 Rns), Log Z from Eqn.1 is plotted against log (sinh $\alpha\sigma$) (a) with the present 55.12 (Q_{HW}= 161 kJ/mol) positioned lower than 7004 with higher Q_{HW} (185 kJ/mol); slopes n_z (0.9, 2.8) and intercepts log A (11.0, 10.7) respectively. As Z declines, the ductility ε_F rises rapidly from 200 to 400°C (b) as increased DRV delays W-cracking [27,28].Variability may result from cast structure.

0.01 s^{-1} (TOR). This graph includes the strengths of 7004 with minor interpolation guided by 7020; however, their T dependencies are less and their $\acute{\varepsilon}$ dependencies greater and more uniform than 55.12 [9,10]. Peak stress versus $\acute{\varepsilon}$ exhibits a steep rise at low rates; when σ_p is plotted against log $\acute{\varepsilon}$ the lines are almost linear (Figure 2b)in agreement with 7004 for 500°C (indicating an exponential law below).

The data was fitted to the constitutive equation [11-16,20,21]:

$$A (\sinh \alpha\, \sigma_p)^n = \acute{\varepsilon} \exp (Q_{HW}/RT) = Z \qquad (1)$$

where A, α (= 0.052 MPa^{-1} common to many Al alloys), n and Q_{HW} are constants (R = 8.31 kJ/mol); Z is the Zener- Holloman parameter that includes the two control variables. In Figure 3a, lines of constant T are fairly straight but the slopes n decrease with T with average 8.03; the results for 7004 are much more consistent with average n= 3.47 similar to the value at 500°C for

Table I. CONSTITUTIVE CONSTANTS (Equation 1) for Al-5.5ZN-1.2Mg (55.12)

Alloy-Mode		α MPa^{-1}	n_{av}	s_{av}	Q_{HW} kJ/mol	n_z	A Log	Reference
55.12	tor-sion.	0.052	1.05	8.03	161	0.9	11.0	Present
as-cast	ten-sion.	0.052	1.75		218			[9]
55.12-tor	as-cast	0.036	3.34		159		9.7	[12]
	homogenized	0.036	3.90		172		11.0	[12]
7004 -tor	revision	0.052	2.79	3.47	185	2.8	10.7	Present
	original	0.052	3.70	3.60	262		15.8	[10]
7075-tor	precipitated	0.052	3.16		148			[30,34]
	solutionized	0.052	2.36		316			[30,34]

the current alloy. The T dependence in Figure 3b shows parallel lines except 0.1 s^{-1} that is much lower but similar to the slopes S of 7004. The Zener-Hollomon parameter is plotted against log (sinh $\alpha\sigma$) (Figure 4a) for the present alloy and for 7004 that has a higher slope (n_z= 2.8) than the present (n_z= 0.9) for which the quality of fit is not as good. The stress exponents for as cast Al-5.5Zn-1.2Mg (in Figure 5b,6) are low and very regular because the gradual cooling of the billet leads to well distributed overaged precipitates that are too big to seriously interact with dislocations; however, the low declines in the flow curves could be related to increased overaging of the precipitates. These varied effects are particularly marked in precipitation hardening alloys tested with controlled precipitate morphologies [21,23].

The power law ($\sigma^{n'} = \acute{\varepsilon}$) is not suitable since n' rises markedly as T drops. The exponential law (exp ($\beta\sigma$) = $\acute{\varepsilon}$) is suitable with β rising from 0.072, 0.068 to 0.113 MPa^{-1} as T drops (avg. 0.084 MPa^{-1}) [9]; lower strength 6060 and 6061 have lower dependence on T but similar on $\acute{\varepsilon}$ [9,25,26]. For 7004, the stress multiplier α in Eqn. 1 was varied between 0.01 and 0.14 MPa^{-1} to find the optimum in comparison to the frequntly used 0.052 MPa^{-1} [15,16,22,23]. Increase of α in plots of log $\acute{\varepsilon}$ versus log sinh $\alpha\sigma$ causes the constant T lines to displace and decrease in slope; the optimum occurs when lines become closest to parallel [9]. As α increases, n decreases almost inversely but S rises almost linearly; consequently Q_{HW} falls slightly to become almost constant from α = 0.04 to 0.08 MPa^{-1} [9]. Such behavior has been found in analyses of many Al alloys including 5083, 5182 and 6060 [15,16,21,25]. The constancy of Q_{HW} with varying α is of some importance since it facilitates comparison of the T dependencies of alloy strengths on basis of composition and microstructure. With α = 0.04 MPa^{-1} for 7004, Q_{HW} is 262 kJ/mol (n = 4.7) [9] much larger than 153 kJ/mol (n = 5.2) for Al (250/500, $\Delta\sigma$ = 66) [19,21].

The fracture strain ε_F generally increases with rising T (at constant $\acute{\varepsilon}$) but with falling $\acute{\varepsilon}$ (Figure 4b), as increasing dynamic recovery (DRV) slows W-cracking by relieving stresses at triple junctions from differential grain boundary sliding; the ductility decline at 1s^{-1} could be due to pore formation active at 500°C in Al [27,28] or to formation of grain boundary precipitates as in 7075 in which the ductility at 0.5 s^{-1} declined from 2.5 to 1.7 across 300 to 400°C [28,29].

The microstructures on tangential sections at about 0.9 radius were examined in polarized light at two magnifications 50x (Figure 5) and 200x (Figure 6). The torsion axis is vertical and the lower edge is the fracture surface that is usually distorted and not in a place normal to the axis. The appearances can be divided into two groups: those near 300°C with low strains (Figure 5a,6a) in which the strains are low so the grains are elongated to an aspect ratio about 2 and are inclined at 30-45° to the axis (increasing with ε_F). The grains exhibit weakly contrasting subgrains deformation bands and slightly serrated boundaries. Microstructures at/above 400°C (Figure

40

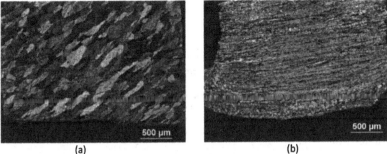

(a)   (b)

Figure 5. Polarized optical micrographs POM of tangential sections at 50x, each with torsion axis vertical and fracture surface at bottom; a) slightly elongated at low T (300°C, 5 s^{-1}, ε_F = 0.2) and b) highly elongated at high T (400°C, 5 s^{-1}, ε_F = 3.2).

(a)   (b)

Figure 6. POM at 200x (oriented as in Figure 6) exhibiting: a) slightly elongated grains with vague subgrains and deformation bands (Figure 6a), b) elongated grains with subgrains and serrations (Figure 6b), having SRX grains near fracture surface (region unloaded before complete failure and removal for quenching)

5b,6b) exhibit grains that are very elongated and thinned except near the fracture surface where there are statically recrystallized (SRX) grains because the irregular progress of the crack permitted $\acute{\varepsilon}$ to fall to zero before complete failure and quenching. As previously observed [7], the grains exhibit well formed subgrains, only a few across each grain, and the strong serrations match the subgrain size. The SRX grains are much larger than the subgrains and exhibit no substructure. It thus appears clear that softening during deformation is dependent on DRV as is common in most Al alloys where there are no large dispersoid particles [14,19,20,30,31].

Discussion

The activation energy for 7004 is rather high because it is in the solution treated condition. There are strong effects of dynamic precipitation at 300°C but reduced effects at 400°C and little at 500°C [23]. The sharp peaks and work softening are similar to flow curves observed for 7075 (Al-5.8Zn-2.5Mg-1.8Cu-0.16Mn-0.18Cr) and 7012 (Al-6.3Zn-1.9Mg-1.1Cu-O.14Mn-0.11Zr) in

41

the solution treated condition as a result of dynamic precipitation followed by particle coalescence [22]. In the pre-precipitated condition, flow curve peaks/softening were much less pronounced (108 MPa compared to 160 MPa for solutionized at 300°C, 1 s^{-1}). Such influences have the following effects in 7075 and 7012: for precipitated the Q_{HW} (n) are 148 (3.16) and 145 (2.94) kJ/mol respectively and for solution treated 300 (2.36) and 272 (2.26) kJ/mol respectively [22]. The Al-Zn-Mg alloys exhibited well recovered DRV substructures with no evidence of DRX; fibrous structure was also observed in extrusions [4-6]. The prior heat treatment of a precipitation hardening alloy has a strong effect on the workability (at 300°C, 1 s^{-1} $\sigma_p \approx$ 108MPa for precipitated and 160MPa for solutioned) [22,23] whereas solute alloys such as Al-Mg are not so affected [17,18,30,31] as discussed below. The much softer 6060 (Al-0.4Mg-0.5Si) in a precipitated condition has lower Q_{HW} = 160kJ/mol whereas the slightly softer 6061 (Al-0.97Mg-0.71Si-0.28Cu) affected by dynamic precipitation has Q_{HW} = 205 kJ/mol. Their low hot strengths are related to an alloying content less than a half those of Al-Zn-Mg alloys [9-13,23,25,26].

The 55.12 alloy is similar to one studied by Ronning et al. [12-14,44] in an extensive project on Al alloys with 4.5-7.5 Zn combined with 0.8 to 1.9 Mg (some with small additions of Cr or Zr). The sinh constitutive constants were n=3.6 and Q_{HW}= 159 kJ/mol for as-cast and n=3.9 and Q_{HW}= 176 kJ/mol for homogenized. The activation energy declined slightly with rising Zn or Mg concentrations [11]. At 450°C, 1 s^{-1} for 1.2 Mg, σ_P rose from 50 to 54 MPa as Zn rose form 4.5 to 7.5% whereas for all Zn levels σ_P rose from 45 to 62 MPa as Mg climbed from 0.8 to 1.8%. While Zn provides valuable strengthening at 20°C, it strengthens little at high T due to negligible solute drag providing valuable workability [30,31] unlike Mg [17,18,30,31]. For all alloys in torsion tests to fixed strains rapidly cooled specimens exhibited no recrystallization. In extrusions (cooled in air) there was often considerable SRX except for alloys with Cr or Zr [11]. In other extrusion studies, rapid cooling or added dispersoid preserved the fibrous grain structure [4-6].

In comparison to Al-Mg alloys, the Al-Zn-Mg alloys have strengths higher than 5005 [1,9] but are much lower than Al-5Mg or 5083, where the declines after the peaks are 50% higher and have been attributed to release of dislocations from solute pinning and enhanced DRV [15], and in the second additionally to particle stimulated nucleation of DRX [14-16,20]. Up to 300°C, ε_F are only slightly better than 5083 whereas at 400°C they are almost a magnitude larger. However, unlike Al-Mg alloys and similar to other 7000 series alloys, ε_F decreases at 500°C due to the formation of large precipitates at grain boundaries GB [30,34]. From 300 to 500°C, the 1 s^{-1} strength declines from 117 to 41 MPa ($\Delta\sigma_P$ =76MPa) whereas the σ_P range for 0.1-10 s^{-1} remains constant at about 28 MPa so the fractional range varies from 25 to 70%. In comparison 5083 has about ~70% higher strength for 1 s^{-1} at both T ($\Delta\sigma_P$ = 95MPa) and the $\Delta\dot{\varepsilon}$ range for σ_P rises in magnitude as T rises so the fractional variation is 30 to 90% [1]. The high $\Delta\dot{\varepsilon}$ range in 5083 is related to high strain rate sensitivity and hence low stress exponent n in the constitutive equation.

The Al-Zn-Mg alloys (Zn+Mg $\approx$5-7%) have lower hot strength than alloys with 5%Mg as solute because the Zn in solution has no solute drag effect [30,31] and part is trapped in large precipitates with Mg. The present Q_{HW} is not unusual compared to those for Al-Mg-Mn alloys: for 5005 Q_{HW} = 208 kJ/mol and n = 4.4; for 3004 Q_{HW} = 207 kJ/mol and n = 4.4; for 5182 Q_{HW} = 185 kJ/mol and n = 2.7; for 5083 Q_{HW} = 162 KJ/mol and n = 1.9 [15,16]. The Q_{HW} did not increase in order of increasing alloy content and rising $\Delta\sigma_P$ (250/500°C); while 5083 has the highest S it has a very low n which brings down Q_{HW} = 2.3 RnS [15,16]. The low value of n is related to a high strain rate sensitivity ($\approx$ 1/n) arising from solute drag by Mg atoms. The high values of Q_{HW} partly arise from the impurities having a strong effect on strength at low T [19]. The decrease in Q_{HW} due to Mg solute drag has support from a subgrain structure and ductility in alloys of 2% Mg similar to Al but much larger than in Al-5Mg [14,30].

Conclusions

As temperature rises above 400°C and strain rate declines, the flow curve strain hardens with much diminished peak to a lower steady state with a higher ductility. The hyperbolic sine constitutive equation is suitable and the current results are similar but more variable (as-cast condition) than those reported in the literature. While providing useful room temperature strength, the hot strengthening from Zn is much less than from Mg because of the absence of solute drag. The quenched specimens are fibrous at all conditions and at/above 400°C exhibit elongated grains with serrations and several subgrains across; the first formed edges of the cracks show recrystallized grains.

References

1.   J. E. Hatch, ed. *Aluminum. Properties, Physical Metallurgy*,ASM, Metals Park,OH, 1984.

2.   H. J. McQueen and O. C. Celliers, *Canadian Metal. Quart.*, 36 (1997), 73-86.

3.   K. Laue and H. Stenger, *Extrusion: Process Machinery, Tooling*, American Society for Metals, Metals Park, OH (1981), 1-62; 124-152.

4.   Y. Baba and H. Yoshido, *Proc 2nd Intnl. Al Extrusion Tech. Sem.*, Aluminum Association, Washington (1977), vol. 1, 301-306.

5.   T. Sheppard and D. Raybould, *J. Inst. Met.*, 101 (1973) 33-34, 73-78.

6.   D. Raybould and T. Sheppard, *J. Inst. Met.*, 101 (1973), 45-52, 65-72.

7.   J. Shen, … G. Gottstein, Mat. Sci. Forum, 426-432 [5] (2003), 3843-3848.

8.   P. Leo, E. Cerri, H. J. McQueen and A. Taurino, *Aluminium Alloys, Their Physical Mechanical Propereties ICAA 11,* DGM, Frankfurt, Germany (2008), 1868-1874.

9.   H. J. McQueen and N. Owen, *Innovative Technologies for Steel (G. Heffernan Symp.)*, J. Guerard, E. Essadiqi, eds., Met. Soc. CIM, Montreal (2001), 295-302.

10.   Y.V.R. K. Prasad and X.Y. Sasidhara, *Hot Working Guide, a Compendium of Processing Maps*, ASM Intl., Materials Park, OH (1997), 136-137.

11.   B. Ronning, *Constitutive Relationships for AlZnMg, AlZnMgCr and AlZnMgZr alloys*, D. Ing. Thesis, Norwegian Univ. Sci. Tech. Trondheim, (1998).

12.   B. Ronning and N. Ryum, *Phys. Met. Mat. Sci.*, 32A (2001), 769-776.

13.   M. El Mehtedi et al., *Metal. Sci. Techn.*, 22 (2004), 3-8.

14.   H. J. McQueen et al., *Met. Sci.*, 18 (1984), 395-402.

15.   H. J. McQueen and J. Belling, *J. Mat. Eng. Performance*, 10 (2001), 164-172.

16. H. J. McQueen and J. Belling, *Can. Metal. Quart.*, (2000), 483-492.

17. H. J. McQueen and W. Blum, *Aluminium 80* (2004), 1151-1159, 1263-1270, 1347-1355.

18. W. Blum, Q. Zhu, R. Merkel and H. J. McQueen, Z. Metalkde, 87 (1996), 14-23.

19. H. J. McQueen and N. Ryum, *Scand. J. Met.*, 14 (1985), 183-194.

20. T. Sheppard et al., *Microstructural Control in Al Alloy Processing*, H. Chia and H. J. McQueen, eds., TMS-AIME, Warrendale, PA (1985), 19-43, 123-154, 155-178.

21. H. J. McQueen and P. Sakaris, *Aluminum Alloys: Their Physical, Mechanical Properties ICAA3*, L. Arnberg et al., eds., NTH- Sinteff, Trondheim, Norway(1992), vol. 2, 179-184.

22. E. Cerri et al., *Mat. Sci. Eng.*, A197 (1995), 181-198.

23. H. J. McQueen, *Hot Deformation of Aluminum Alloys*, T. G. Langdon and H. D. Merchant, eds., TMS-AIME, Warrendale, PA (1991), 105- 120.

24. H. J. McQueen, *Aerospace Materials and Manufacturing IV: Advances in Processing/Repair*, M. Jahazi, et al. eds., Met. Soc., CIM, Montreal (2008), 111-123.

25. H. J. McQueen and M. J. Lee, *Al Alloys, Physical and Mechanical Properties (ICAA7)*, E.A. Starke and T. Sanders eds, TransTech Pub., Zurich (2000), 437-442.

26. E. Herba and H. J. McQueen, *Hot Workability of Steels and Light Alloys-Composites*, H. J. McQueen et al. eds., Met. Soc. CIM, Montreal (1996), 53-60.

27. M. E. Kassner et al. , *Mat. Sci. Eng.*, A132 (1991), 97-105.

28. H. J. McQueen, *Recent Developments in Processing/Applications of Structural Alloys*, S. Spigarelli, M. Cabibbo, eds., *Key Eng. Mat.* (2009), in press.

29. H. J. McQueen, E. V. Konopleva and G. Avramovic-Cingara, *Proc. 11th Intnl. Conf. Composite Materials*, M.L. Scott, et al. eds., Australian Comp. Struc. Soc., Melbourne (1997), vol. III, 418-428.

30. H. J. McQueen, W. Blum, Q. Zhu and V. Demuth, *Advances in Hot Deformation Textures and Microstructures*, J. J. Jonas et al. eds., TMS-AIME, Warrendale, PA (1994), 235-250.

31. Q. Zhu, V. Demuth, W. Blum and H.J. McQueen, Strength of Materials (ICSMA/10, Sendai), H. Oikawa et al., eds., Japan Inst. Metals (1994), 803-806.

Aluminum Alloys: Fabrication, Characterization and Applications II
Edited by: Weimin Yin, Subodh K. Das, and Zhengdong Long
TMS (The Minerals, Metals & Materials Society), 2009

MICROSTRUCTURAL CONTROL THROUGH HEAT TREATMENT PROCESS IN AN AEROSPACE ALUMINUM ALLOY

Zainul Huda

Department of Mechanical Engineering
University of Malaya
50603 Kuala Lumpur
Malaysia

Keywords: 2017 aluminum alloy, precipitation strengthening, age hardening, microstructure

Abstract

The 2017 aerospace aluminum alloy has been characterized through metallographic investigations. A series of precipitation strengthening and age-hardening heat treatment processes involving solution treatment at 550 °C followed by quenching (and tempering for various time-durations) were conducted for the 2017 alloy. Microstructural characterization of the heat-treated samples showed effective distribution of fine θ' particles in the α-matrix of the aluminum alloy; these microstructural features enable us to develop proper precipitation strengthening and age-hardening heat-treatment process parameters for the 2017 aluminum alloy for aerospace application.

Introduction

The 2xxx series age-hardenable aluminum alloy are extensively used in aircraft structures owing to their good specific strength and lightweight [1-3]. The 2017 aluminum alloy, in the as-rolled condition, is unsuitable for aerospace application since it lacks strength and ductility owing to elongated grains, regions of high energy, and absence of dispersed second-phase particles in its microstructure. These microstructural features require proper precipitation-strengthening or age-hardening heat treatment to be given to the alloy for aerospace application [4-5].

The metallurgically important feature of 2xxx series aluminum alloys is the ability to improve mechanical properties when suitably heat treated. Aluminum-copper alloys (for aerospace applications) are usually given special heat treatments, called age-hardening which are process of strengthening metals based on θ'-particles strengthening. For the process to occur, it requires certain phase transformations resulting from either precipitation strengthening or age hardening heat treatment involving solution treatment, quenching and tempering [6].

The aluminum-rich portion of Al-Cu equilibrium phase diagram enables us to determine solution-treatment temperature for age hardening of the Al-Cu aerospace aluminum alloy. If the Al-Cu alloy (containing less than 5.7%Cu) is slowly heated at above-solvus temperature (and below liquidus temperature), the particles of $CuAl_2$ are absorbed until we obtain a single-phase solid solution comprising of α-phase. On quenching the alloy, we retain the copper in solution, and in fact, produce a supersaturated solution of copper in aluminum at room temperature. If the quenched alloy is allowed to remain at room temperature, it is found that strength and hardness gradually increases and reaches a maximum in several days. The completely α-phase structure obtained by quenching is not the equilibrium structure at room temperature. It is in fact supersaturated with copper, so copper atoms diffuse out according to the following phase transformation:

$$Cu \quad + \quad 2Al \quad \longrightarrow \quad \theta' \qquad (1)$$
$$\text{[Al lattice]} \quad \text{[Al lattice]} \qquad \text{[Intermediate coherent precipitates of } CuAl_2\text{]}$$

The precipitation of θ' (see Equ 1) in the microstructure of the 2xxx series Al-Cu alloy greatly increases strength renders the material suitable for aerospace applications [7].
The objective of the research reported in the paper aims at developing a suitable precipitation strengthening and age-hardening heat treatment process with specified parameters for the 2017 alloy. An attempt is also made to study the effect of solution treatment followed by slow cooling and the effect of overheating on solution-treated and quenched alloy on its microstructure and hardness.

Experimentation

The starting material (SM), an 2017 aluminum alloy plate, for the research was acquired from local market. The chemical composition of the 2017 aluminum alloy is presented in Table 1.

Table 1: chemical composition of the 2017 aluminum alloy

Cu	Si	Fe	Mn	Mg	Zn	Cr	Ti	Al
4.3	0.4	0.7	0.6	0.6	0.25	0.1	0.15	balance

Five (5) metallographic samples were sectioned from the SM by use of a scissor. Four (4) out of the 5 samples were heat treated in an atmosphere-controlled furnace according to the sample-identification scheme presented in Table 2.

Table 2: Sample identification scheme according to heat treatment

Heat Treatment Process	Sample Id #
As-received material	A-0
Heated to 550 °C then slowly cooled	B
Heated to 550 °C, quenched, aged for 2 days	C
Heated to 550 °C, quenched, tempered 165 °C/10 h	D
Heated to 550 °C, quenched, tempered 200 °C/10 h	E

The five (5) samples (see Table 2) were metallographically mounted by cold mounting technique. Metallographic specimens for the 5 samples were prepared by metallographic grinding (with 800, 1000, 1200, and 1500 grit size silicon carbide emery papers) and polishing (using high alumina powder) followed by metallographic etching by use of Keller's reagent (150 ml water, 3 ml HNO_3, 6 ml HCl, and 6 ml HF). Microstructural characterization involved use of an optical microscope linked with a computerized imaging system equipped with MSQ software. The sophisticated camera system installed with the optical microscope enables us to capture images of various samples (see Table 2). A Vickers hardness testing machine was used to determine hardness values before and after heat treatment practices.

Results

Microscopic Results

The microstructures of the 2017 aluminum alloy before and after heat treatment (see Table 2) are shown in the optical micrographs in Fig 1(a-e).

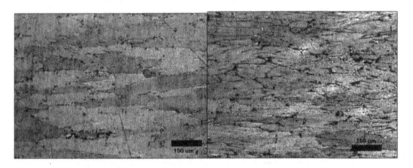

Fig 1(a) Microstructure of the as-received aluminum alloy (sample A-0)   Fig 1(b) Microstructure of aluminum alloy heated to 550 °C then slowly cooled (sample B)

Fig 1(c) Microstructure of age-hardened aluminum alloy heated to 550 °C, water-quenched then cooled at room temp for 2 days (sample C)   Fig 1(d) Microstructure of precipitation-strengthened aluminum alloy heated to 550 °C, water-quenched then tempered 165 °C/10 h (sample D)

47

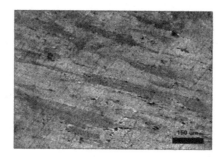

Fig 1(e) Microstructure of aluminum alloy heated to 550 °C, water-quenched then tempered at 200 °C/10 h (sample E)

Hardness Test Results

The results from Vickers hardness testing machine for the samples (A-E) (see Table 2) are presented as a column chart in Fig 2.

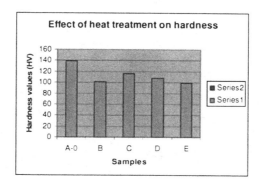

Fig 2: Column chart showing effects of heat treatment on hardness of 2017 aluminum alloy

Analysis and Discussion

Effect of solution treatment followed by slow cooling

Prior to discussing effect of heat treatment on the investigated material (2017 aluminum alloy), we first analyze microstructure and hardness test results for the as-received material. The microstructure of the as-received 2017 aluminum alloy is shown in Fig 1(a). The interpretation of the microstructure in Fig 1(a) reveals elongated grains indicating that the starting material was in the un-recrystallized form; the microstructure indicates high-energy regions resulting from cold rolling. This fact is also confirmed by

48

the high hardness value (139 HV) of the as-received materials (see Fig 2). Hence, it is quite logical to conclude that the as-received material lacks ductility and is unsuitable for aerospace application.

Now we discuss the effect of solution treatment followed by slow cooling given to the as-received material. The microstructure of the heat treated alloy (sample B) (see Table 2) is shown in Figure 1(b); which shows precipitation of θ phase along grain boundaries in the microstructure of sample B. The θ phase (in the microstructure of Sample D) is intermetallic compound ($CuAl_2$) which forms as non-coherent precipitate. The presence of non-coherent particles/phase of $CuAl_2$ will cause the material to become weak because only 0.2% Cu is left in solution [7-8]. This microstructural interpretation is in accordance with the fall in hardness value of sample B as compared to sample A-0 (hardness value falls from 139 to 101 HV) (see Fig 2). Hence, it is justified to conclude that the alloy resulting from solution treatment at 550 °C followed by slow cooling to room temperature lacks strength and is also unsuitable for aerospace application.

Effects of Age-hardening (Natural Aging)
The effect of age hardening heat treatment (involving solution treatment at 550 °C, quenching and then cooling at room temperature (30 °C) for 2 days) on the microstructure of the 2017 aluminum alloy is shown in the micrograph in Fig 1(c). It is clearly observed that a second phase ($CuAl_2$) (θ') has diffused out from the earlier supersaturated state after aging for 2 days. Since the supersaturated state resulting from quenching is not the equilibrium state, the copper atoms diffuse out from aluminum lattice and form a coherent θ' and the fine dispersion of precipitates within the grains (see Fig 1(c)). The strengthening imparted by the dispersed precipitates in the age-hardened alloy (sample C) can be verified by the rise in hardness value from 101 to 116 HV (see Fig 2). These microstructural features and hardness test analysis lead us to a conclusion that the strength of the naturally-aged alloy is good enough for aerospace application [7-9].

Effects of Precipitation Strengthening (Artificial Ageing)
Figure 1(d) shows microstructure of precipitation-strengthened aluminum alloy (sample D); here a second phase also precipitates as the intermediate coherent precipitates of $CuAl_2$ (θ'). The fine particles of θ' phase effectively impede the movement of dislocations and hence impart strength to the alloy [5-10]. An analysis of the column chart shown in Fig 2 indicates a slightly lower hardness value of sample D (108 HV) in comparison to sample C (116 HV) (see Fig 2); however this slightly lower hardness value does not necessarily indicate lower strength. In fact, microstructure of sample D (see Fig 1(d)) shows quite effective dispersion of θ' particles; and hence advocates both good strength as well as better ductility. This micrsotructural interpretation is in agreement with literature; according to which a high tensile strength of 470 MPa was imparted in a similarly-treated Al-4%Cu alloy [8]. It is, therefore, quite logical to conclude that the artificial ageing heat treatment given to the alloy (sample D) is the optimum precipitation strengthening parameters to render the material suitable for aerospace application.

Effect of Overheating tempering on quenched alloy

49

The effect of overheating tempering on the solution-treated and quenched alloy is shown in the micrograph in Fig 1(e). The microstructure of sample E reveals coarse particle phase; it means tempering at a higher temperature has caused precipitation of coarse θ'' phase which is non-coherent to matrix [7]. This microstructural feature indicates loss in strength; which is confirmed by a low hardness value (99 HV) for the sample E (see Fig 2). The material behavior leads us to draw the conclusion that the strength of the alloy (sample E) is poor; and hence the material is unsuitable for aerospace application.

Conclusions

The metallographic and hardness testing investigations for 2017 aerospace aluminum alloy lead us to draw the following conclusions:
a) The microstructure of 2017 aluminum alloy resulting from solution treatment at 550 °C followed by slow cooling to room temperature indicates low strength and is unsuitable for aerospace application.
b) Natural ageing (age hardening) involving solution treatment at 550 °C, water-quenching followed by ageing at room temperature (30 °C) for 2 days (48 h) imparts adequate strength to the alloy; and renders the material suitable for aerospace application.
c) Artificial ageing (precipitation strengthening) involving solution treatment at 550 °C, water-quenching followed by tempering at 165 °C for 10 hours results in effective dispersion of fine θ'-phase particles in the microstructure; and was found to be the optimum precipitation strengthening parameters to render the material suitable for aerospace application.

References

1. İbrahim Özbek; A study on the re-solution heat treatment of AA 2618 aluminum alloy
   Materials Characterization, Volume 58, Issue 3, March 2007, Pages 312-317
2. C. Y. Xie, R. Schaller, C. Jaquerod; High damping capacity after precipitation in some commercial aluminum alloys
Materials Science and Engineering A, Volume 252, Issue 1, 31 August 1998, Pages 78-84
3. Z. Huda, M. Saufi and Shaifulazuar; Mechanism of Grain Growth in an Aerospace Aluminum Alloy; Journal of Industrial Technology; 15 (2) 2006, 127-136
4. H. Kacer, E. Atik, and C. Meric; Journal of Materials Processing Technology, Vol. 142, Issue 3, 2003, 762-766
5. P. DeGarmo, J.T. Black, and R.A. Kohser; Materials and Processes in Manufacturing; John Wiley Publications, 2003
6. V.B. John; Engineering Materials, MacMillan Press, London, 1990
7. D.R. Askland and P.P. Phule; The Science and Engineering of Materials, Thomson Books Inc USA, 2003
8. R.A. Higgins, Engineering Metallurgy, Edward Arnold Publishers, London, 1980
9. F.W. Gayle and M. Goodway, *Precipitation hardening in First Aerospace Aluminum Alloy: The Wright Flyer Crankcase;* **Science,** 1994. 266(5187); 1015-1017
10. E. Bishop, Metallurgy of Aluminum Alloys, Chapman and Hall, London, 1967

Aluminum Alloys: Fabrication, Characterization and Applications II
Edited by: Weimin Yin, Subodh K. Das, and Zhengdong Long
TMS (The Minerals, Metals & Materials Society), 2009

INFLUENCE OF Ge-Si ADDITIONS ON AGEING RESPONSE AND PROPERTY IMPROVEMENT IN 2XXX ALLOYS

L. Zhuang[1], S. Chen[1], A.F. Norman[1], A. Bürger[2], and S. Spangel[2]

[1]Corus Research, Development & Technology,
P.O. Box 10000, 1970 CA IJmuiden, The Netherlands
[2]Aleris Aluminum Koblenz GmbH, Carl-Spaeter-Straße 10,
56070 Koblenz, Germany

Keywords: Al-Cu alloy, Ge-Si clusters, ageing kinetics, strength, fracture toughness

Abstract

2xxx alloys containing additions of Ge-Si display a unique combination of rapid ageing response, higher peak strength, and extended thermal structure stability. Furthermore, due to a much refined distribution of precipitation, 2xxx + Ge/Si alloys show a considerable improvement in the tensile yield strength while retaining good fracture toughness, or damage tolerance. High resolution transmission electron microscopy (HRTEM) examination showed that the Ge-Si additions modify the conventional precipitation sequence in Al-Cu alloys. In Ge-Si added alloys, Ge-Si clusters quickly nucleate and grow during elevated-temperature ageing. The Ge-Si particles then act as nucleation sites for θ' precipitates, resulting in a peak-aged microstructure consisting of a dense distribution of θ' attached to fine Ge-Si particles. A similar effect was also observed in Ge-Si/Mg added alloys.

Introduction

The essential factors that control the properties of precipitation hardened structural alloys are the type, size and distribution of the strengthening precipitates in the metal matrix. The elements Si, Mg, Be, Ge, In, and Cd have been proved to modify the dominant precipitation reaction in Al-Cu based alloys [1-2]. One mechanism by which the precipitation reaction is altered is the preferential precipitation of alloy modifiers, producing heterogeneous sites for the formation of θ' phase in Al-Cu alloys [3-5].

In Al-Si-Ge alloys, much finer Ge-Si precipitates can be obtained than in Al-Si or Al-Ge binary alloys [4-6]. The Ge-Si precipitates form a diamond cubic structure due to the compromise of the Si and Ge elements (the atomic diameter of Si is smaller than that of Al but the atomic diameter of Ge is larger). Because of the cancellation of the misfit stresses, Ge-Si precipitates do not contribute strength remarkably, and therefore the strength of Al-Ge-Si alloys remain relatively low. It is also reported that the ageing kinetics of Al-Ge-Si alloys is surprisingly slow [4]. However, the simultaneous addition of Ge and Si to Al-Cu alloys modifies the precipitation significantly [7-8]. Not only are the ageing kinetics accelerated but also the peak ageing strength is increased.

The objective of this investigation is to understand the effect of alloying Ge, Si in Al-Cu based alloys, and to extend this concept to examine the effect of forming Si/Ge-Mg clusters on the precipitation behavior and the mechanical properties in Al-Cu alloys. It will be shown that the alloys with Ge, Si and Mg additions showed the similar effect on accelerating the ageing kinetics observed in Al-Cu-Ge-Si alloys. However these alloys give surprisingly much stronger strengthening while retain good fracture toughness.

Experimental

A number of Al-Cu based alloys with additions of Ge, Si and/or Mg were cast. Alloy 1 was the reference alloy, and for alloys 2 and 5 the concentration of Si and Ge (Mg) were kept stoichiometric. Alloys 3 and 4 can be treated as removing Ge or Si in the chemistry of alloy 5. The compositions of the main elements of the investigated alloys, as analyzed in as-cast ingots, were listed in table 1. The actual compositions have a slight deviation from the designed compositions.

Table 1 The chemical compositions of investigated alloys (in wt.%)

Alloy	Compositions			
	Cu	Ge	Si	Mg
1	4.5	-	0.05	-
2	4.6	0.69	0.26	-
3	4.6	0.69	-	0.27
4	4.6	-	0.41	0.32
5	4.6	0.67	0.41	0.33

The cast ingots were homogenized, hot rolled and cold rolled to sheets of 2 mm in thickness. The sheet materials were solution heat treated, water quenched and artificially aged. The solution heat treatment temperature was 510°C for all the alloys. The ageing temperature for T6 treatment was 190°C.

Brinell macro-hardness was measured using B scale to establish ageing curves and the focus was looking at the ageing kinetics. DSC analysis and (HR)TEM were applied to characterize the ageing behavior. Tensile and Kahn tear testing (following ASTM B871-01) were conducted to evaluate the mechanical properties.

Results and discussion

Alloying effect on ageing kinetics and hardening

Ge-Si or Mg containing alloys show a similar natural ageing behavior as the reference Al-Cu alloy. However, the ageing kinetics during artificial ageing treatment is significantly changed.

Figure 1 shows the ageing curves at 190°C for various alloys after solution heat treatment at 510°C, water quenching, and natural ageing for 2 weeks. As shown that much faster ageing kinetics are achieved in Al-Cu based alloys with Si-Ge/Mg additions. The ageing time at 190°C to reach the peak strength for the Al-Cu alloy is about 12 hours, whilst the peak-ageing time is substantially reduced to about 3 hours in the Al-Cu-Si-Ge/Mg alloys.

For the reference Al-Cu alloy, a hardness decrease is shown at beginning of artificial ageing. This is attributed to the partial dissolution of GP zones or clusters, which are formed during natural ageing. These GP zones or clusters are not stable (either dimensional or chemical) upon heating during artificial ageing. Furthermore, the dissolution process leads to a long incubation time before effective hardening can occur.

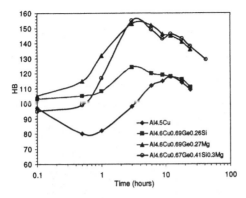

Fig. 1. The ageing curves of Al-Cu based alloys showing the effect of Ge-Si/Mg additions on the ageing kinetics.

However, in the Ge-Si/Mg containing alloys no softening stage is observed, indicating that the dissolution process of non-stable GP zones or clusters has been largely eliminated. The alloy with the combined addition of Ge, Si and Mg exhibits the most accelerated ageing kinetics.

No substantial hardening has been achieved in Al-Cu-Ge-Si alloy as compared to that in Al-Cu reference alloy. This is due to the cancellation of the misfit stresses as reported in previously publications [4-6]. However, with the extra addition of Mg or by replacing Ge/Si by Mg, a significant increase in peak hardness is observed.

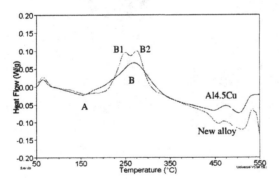

Fig. 2. DSC curves of the reference Al-Cu alloy and the new alloy with Ge-Si/Mg addition in the T4 condition.

In addition, it should be noted that the hardness decreases very slow after peak-ageing is reached in Ge-Si/Mg containing alloys. This suggests an extended thermal stability of the precipitate structure in these alloys.

53

The accelerated ageing kinetics observed in the Ge-Si/Mg added alloys can be explained by the DSC profiles given in Fig. 2. First of all, the new alloys showed no clear dissolution reaction in the DSC profile, in line with the ageing curves given in Fig. 1. In the Al-Cu alloy, only one exothermic peak "B" is shown upon heating from its natural aged condition which can be related to the formation of the hardening precipitation of the θ' phase. In the new alloys, the exothermic peak "B" is split into two peaks. Peak "B1" may be related to the formation of Ge-Si/Mg clusters, whereas "B2" corresponds to the formation of the θ' phase.

Fig. 3. Typical precipitate microstructure in the peak aged Al-4.5Cu alloys.

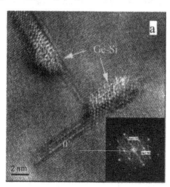 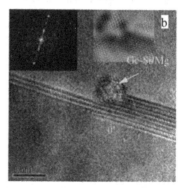

Fig. 4. The high-resolution TEM images of the microstructure in the peak aged samples. (a) Al4.6Cu0.69Ge0.26Si, (b) Al4.6Cu0.67Ge0.41Si0.33Mg.

TEM examination confirms the formation of Ge-Si clusters or Ge-Si/Mg clusters in those new alloys before the θ' precipitates are formed. Figure 3 shows TEM images from the peak-aged samples of the reference Al-Cu alloy. A typical needle-like precipitation structure is observed in this alloy in the peak-aged condition.

However, as shown in Fig. 4, in the Ge-Si or Ge-Si-Mg containing alloys, the Ge-Si or Ge-Si/Mg clusters were formed in the order of a few nanometers in size. These clusters serve as nucleation sites for the θ' precipitates. As a result, this accelerates the entire ageing kinetics in these alloys. Furthermore, the pre-existing clusters lead to a much finer θ' precipitation in these alloys than was previously observed in the Al-Cu alloy. A much refined distribution of precipitation can provide a considerable improvement in fracture toughness, or damage tolerance in high strength alloys.

Mechanical properties

Fig. 5 ohowo tho relationships between the tensile yield strength and the UPE values and TS/Rp of the alloys under investigation. The UPE values can be used as an indication of the resistance to crack growth rate of the alloys, while the TS/Rp values as an indication of fracture toughness.

In general, for Al-Cu based alloys, an increase in strength leads to a corresponding decrease in the UPE values. For example, when the Cu content is increased from 4.5 wt.% to 5.7 wt.%, the yield strength is increased from about 235 MPa to about 295 MPa, while the UPE value decreased from about 250 kJ/m^2 to 150 kJ/m^2.

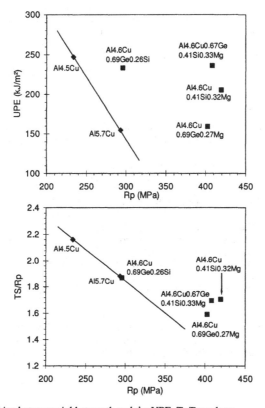

Fig. 5 Relationships between yield strength and the UPE, Ts/Rp values.

55

For the Al4.6Cu alloy with additions of 0.69Ge and 0.26Si, the alloy can reach about the same level of strength as that of Al-5.7Cu, but retain the high UPE value as that of the Al-4.5Cu reference alloy due to a much refiner distribution of precipitates at the peak aged state.

Alloying with Mg to replace Ge or Si leads to a much higher strength which is at the expense of lower UPE values referring to the Al4.6Cu0.69Ge0.26Si alloy. However, the combined addition of Ge, Si and Mg gave the best combination of the properties. A combination of high strength, above 400 MPa, and high UPE value, about 235 kJ/m^2 is achieved in the Al4.6Cu0.67Ge0.41Si0.33Mg alloy.

Additions with Ge-Si/Mg in Al-Cu alloys improve also the fracture toughness, Ts/Rp values when the same yield strength is used as reference. However, the effect is less significant as compared with that on the UPE values.

Conclusions

1. A rapid ageing response, and high peak strength is achieved by adding Ge, Si/Mg in Al-Cu based alloys. The accelerated ageing kinetics results from the formation of stable Ge-Si/Mg clusters that can act as nucleation sites for θ' precipitates.
2. Due to a much refined distribution of precipitation, Ge-Si/Mg containing alloys show a considerable improvement in the tensile yield strength while retain high damage tolerance.
3. The AlCuSiGeMg alloy shows the best combination of the properties: quick ageing response, high strength, and excellent damage tolerance.

References

1. J.M. Silcock and H.M. Flower, "Comments on a Comparison of Early and Recent Work on the Effect of Trace Additions of Cd, In, or Sn on Nucleation and Growth of θ in Al-Cu Alloys", Scripta Materialia, Vol. 46, 2002, pp389-394.
2. T.S. Bondan, I.J. Polmear, S.P. Ringer, Precipitation processes in Al-4Cu-(Mg,Cd) (wt.%) alloys: Materials Science Forum, Vol. 396-402, 2002, pp613-618.
3. E. Hornbogen, A.K. Mukhopadhyay, Jr E. A. Starke, Nucleation of the diamond phase in aluminium-solid solutions: Journal of Materials Science, Vol. 28, 1993, pp3670-3674.
4. E. Hornbogen, A.K. Mukhopadhyay, Jr E. A. Starke, Precipitation hardening of Al-(Si,Ge) alloys: Scripta Materialia, Vol. 27, 1992, pp733-738.
5. D. Mitlin, V. Radmilovic, J. W. Morris, Catalyzed precipitation in Al-Cu-Si: Metall.Mater.Trans., Vol. 31A, 2000, pp2697-2711.
6. D. Mitlin, U. Dahmen, V. Radmilovic, J. W. Morris, Precipitation and hardening in Al-Si-Ge: Materials Sciences& Engineering A, Vol. 301, 2001, pp231-236.
7. D. Mitlin, V. Radmilovic, J. W. Morris, U. Dahmen, On the influence of Si-Ge additions on the ageing response of Al-Cu: Metallurgical and Materials Transactions A, Vol. 34A, 2003, pp735-741.
8. V. Maksimovic, Slavicazec, V. Radmilovic, Milan T. Jovanovic, The effect of microalloying with Si and Ge on microstructure and hardness of a commercial aluminum alloy: J. Serb. Chem. Soc., Vol. 68, 2003, pp893-901.

Aluminum Alloys: Fabrication, Characterization and Applications II
Edited by: Weimin Yin, Subodh K. Das, and Zhengdong Long
TMS (The Minerals, Metals & Materials Society), 2009

MICROSTRUCTURE AND MECHANICAL PROPERTIES OF CAST HYPEREUTECTIC Al-Si ALLOYS WITH HIGH MAGNESIUM CONTENT

A. Mandal and M.M. Makhlouf

Advanced Casting Research Center
Worcester Polytechnic Institute, Worcester, MA 01609 USA

Keywords: Mg_2Si, hypereutectic Al-Si alloy, mechanical properties

Abstract

Magnesium in excess of the quantities typically found in commercial hypereutectic Al-Si alloys can produce alloys with enhanced microstructure and attractive mechanical properties. With addition of Mg to hypereutectic Al-Si alloys, the primary silicon phase is suppressed and is replaced with a fine dispersion of small Si particles. However, an abundance of large Mg_2Si particles with Chinese script morphology also forms in the microstructure and unfavorably influences the tensile properties of the alloy. Efforts were made to overcome the negative effects of these particles by manipulating their size and morphology. Several additives were made to a hypereutectic Al-Si-Mg alloy and their effect on the cast alloy was determined. The alloy treated with Misch Metal and Strontium showed promising results. The Mg_2Si particles that formed in castings made from this alloy were very small and almost spherical; and the room temperature tensile and yield strengths of cast bars were remarkably high.

Introduction

Cast hypereutectic Al-Si alloys have low-density, good high temperature mechanical properties, and are wear-resistant; therefore they are widely used in automotive engine components. However, in recent years, yet better performance has been required from these alloys in order to meet the ever-increasing demands for higher output power from engines. Today, almost all aluminum engine blocks are produced by die-casting 390 (Al-17%Si-4.5%Cu-0.5%Mg) alloy [1], and a review of the open literature shows that only very little work has been done recently on developing new hypereutectic Al-Si alloys [2-3]. The literature review also reveals that most of the research effort in hypereutectic Al-Si alloys has been confined to alloys with less than 0.6 wt% Mg. However, it has been recently shown that magnesium contents in excess of the typical quantities found in commercial hypereutectic Al-Si alloys can produce alloys with enhanced microstructures, which may have attractive mechanical properties. In the present study the effects of adding various modifiers on the morphology of Mg_2Si particles in high Mg hypereutectic Al-Si alloys have been investigated.

Materials and Procedures

Hypereutectic Al-14Si-0.2Fe-*x*Mg alloys used in the present work were produced from pure Al ingots, Al-50%Si, Al-52%Mg and Al-80%Fe master alloys. The alloys' composition was measured using spark emission spectrometry (Table I). Each alloy was prepared in a box type resistance furnace and poured into sand cups. All samples for microstructure analysis were extracted from centre of the casting to ensure similar cooling rates and prepared using standard metallographic techniques.

Table I. Composition of Al-14Si-0.2Fe-*x*Mg alloys used in this work.

Alloy	Si	Fe	Mg	Al
A0	14.05	0.23	-	Bal.
A1	13.95	0.22	1.08	Bal.
A2	13.98	0.23	2.05	Bal.
A3	14.01	0.21	3.06	Bal.
A4	14.06	0.25	4.01	Bal.
A5	14.04	0.25	5.02	Bal.

In order to make tensile test bars, the charge materials were melted at 750-800°C in clean boron nitride coated silicon carbide crucibles in an induction furnace. The desired melt composition was checked using a Spectro-max spark test analyzer and then degassed using a rotary degasser for 40 minutes. Standard tensile test specimens were produced in a permanent mold (as per ASTM E8-04) preheated and maintained at 425±5°C during pouring. Tensile tests were conducted at room temperature in an Instron Universal Tensile Testing Machine. The Rockwell Hardness Scale B was used for hardness measurements.

The microstructures were analyzed using optical and scanning electron microscopy, and particle size analysis was performed using an image analyzer. All the samples were etched with Keller's reagent for optical microscopy and image analysis. For SEM studies, the samples were deep etched with 2% HF solution.

Results and Discussion

Establishing the Base Alloy

Figure 1 shows isopleths of hypereutectic Al-14Si-0.2Fe-*x*Mg alloys created using the commercial thermodynamics software Pandat and show that with the addition of Mg, the Al-Si phase diagram is no longer a simple eutectic invariant, but rather many phases form during solidification of the alloys. Most importantly is that Mg_2Si particles form and their amount increases with increased Mg. Mg_2Si may be desirable in aluminum alloys because of its high melting temperature, low density, high hardness, low coefficient of thermal expansion, and reasonably high elastic modulus [4]. However, its presence in the form of large Chinese script particles may detract from the alloys' mechanical properties.

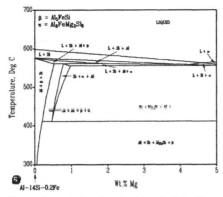

Figure 1. Isopleths of Al-14Si-0.2Fe-xMg alloys

Figure 2 shows typical microstructures of Al-14Si-0.2Fe-xMg alloys solidified in sand molds at 45 C/min. Figure 2a shows the typical microstructure of hypereutectic Al-14Si-0.2Fe-0Mg which contains mainly eutectic silicon together with primary silicon particles dispersed in an α-Al matrix. Addition of 1 wt% Mg to this alloy leads to the formation of Mg_2Si particles with a typical Chinese script morphology. Further addition of Mg (2 wt%) leads to the formation of coarser Mg_2Si particles without any change in their morphology (Figure 2b). In the Al-14Si-0.2Fe-3Mg alloy (Figure 2c), the eutectic silicon is refined and the size of Mg_2Si particles does not increase appreciably. The alloy with 4 wt% Mg does not show any significant change in the size of Mg_2Si particles with respect to the alloy containing 3 wt% Mg. Also the extent of refinement of the eutectic silicon is somewhat diminished. In the alloy with 5 wt% Mg, Mg_2Si crystals (~40 μm) with sharp corners appear in the microstructure along with coarse Chinese script Mg_2Si particles which affect the strength and ductility of the alloy adversely.

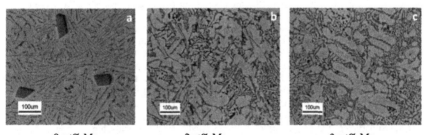

0 wt% Mg   2wt% Mg   3 wt% Mg

Figure 2. Microstructures of Al-14Si-0.2Fe-xMg alloys

Effect of Cooling Rate – Al-14Si-0.2Fe-xMg alloys were cast in a steel wedge mold and a copper wedge mold in order to obtain microstructures under different cooling rates. The cooling rates were calculated to be 220 C/min and 300 C/min for the steel and copper molds, respectively. Although there is a noticeable decrease in the size of Mg_2Si paticles at the higher

59

cooling rate, there is no significant change in its morphology - the Chinese script still persists. At slower cooling rates, the variation in size of the Mg_2Si particles is high, but when the alloys are cooled at relatively higher cooling rates, there is hardly any variation in the size of the Mg_2Si particles with increased Mg content.

Modification of Mg_2Si Morphology by Chemical Additives

The base alloy, namely Al-14Si-0.2Fe-3Mg, was treated with different chemical modifiers (refer Table II) that were believed can alter the Chinese script morphology of Mg_2Si into a more useful morphology. The elements chosen for modification and their addition levels were based on previous studies on Mg based alloys [4-5]. The choice of Ti, Zr and Sn stems from the fact that these alloying elements form particles which have cubic crystal structure and Mg_2Si has fcc structure. This could possibly lead to easier nucleation of Mg_2Si particles and hence a finer microstructure could be obtained. Misch metal is known to refine Mg_2Si particles in Al alloys [6].

Table II. Master alloys used to modify the morphology of Mg_2Si particles.

Master Alloy	Addition level (wt%)
Al-87% Y/Al-99%Ce	0.25, 0.50, 0.75, and 1.0
Al-10wt%Sr	0.025, 0.050, 0.5, and 1.0
Cu-15wt%P	0.02, 0.05, 0.2, and 0.5
Al-10wt%Ca/Al-5.7% Ti/Pure Sn	0.2, 0.5, and 1.0
Pure Zr	0.02, 0.50, and 1.0
Misch Metal	1, 2, and 3

Examination of the resulting microstructures showed that the morphology and size of Mg_2Si particles were not significantly affected by Y, P, Ca, Ti, Zr, and Sn additions. However addition of Ce, Sr, and misch metal affected the morphology of Mg_2Si particles as well as the eutectic silicon particles. Cerium has a pronounced effect on the size distribution of coarse Mg_2Si particles. The typical Chinese script morphology of Mg_2Si is modified and/or fragmented to smaller Mg_2Si particles. This fragmentation is seen in the case of the alloys modified with 0.25 and 0.50 wt% Ce, however the beneficial effect of Ce seems to fade at higher addition levels. Strontium has a similar effect on the morphology of the Mg_2Si particles as cerium, though the effect occurs at much lower Sr levels. At 0.025 and 0.050 wt%, the eutectic silicon and Mg_2Si particles are significantly modified. Combined addition of Ce (0.25%) and Sr (0.025%) resulted in a yet more uniform microstructure. This may be due to the different modifying action of each element. While Ce modifies the Mg_2Si particles, Sr modifies both Mg_2Si particles and the eutectic Si particles. The Mg_2Si particles are modified from the typical Chinese script to various other morphologies as seen in Figure 3.

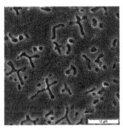

Figure 3. SEM micrograph showing the morphology of Mg_2Si
in Al-14Si-0.2Fe-3Mg-0.25Ce-0.025Sr alloy

Misch metal is a cheaper alternative to Ce and can have a pronounced effect on the morphology of primary Mg_2Si particles. Hence, a systematic study of the effect of misch metal on the hypereutectic base alloy was performed. The amount of misch metal, reaction time, and reaction temperature were chosen as the key parameters in optimizing the microstructure, and it was concluded that a reaction temperature of $800^{\circ}C$ and a reaction time of about 90 minutes were optimum. Also it was found that 1 wt% misch metal was required to refine the coarse morphology of Mg_2Si particles. Figure 4a shows the silicon particles along with Mg_2Si particles and a few rare earth (RE)-containing particles. Figure 4b shows a typical RE particle.

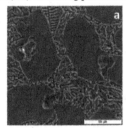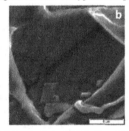

Figure 4. (a) SEM micrograph of Al-14Si-0.2Fe-3Mg-1MM alloy (b) RE particle

Microstructure and Mechanical Properties of Permanent Mold Castings

Figure 5a shows the microstructure of Al-14Si-3Mg-0.2Fe alloy treated with Sr and misch metal

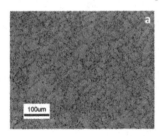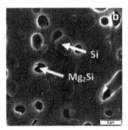

Figure 5. (a) Optical and (b) SEM micrographs of Al-14Si-0.2Fe-3Mg-1MM-0.025Sr alloy

and solutionised at 540°C for 8 hours. The microstructure shows a uniform distribution of spheroidal Si and Mg_2Si particles. The SEM micrograph in Figure 5b clearly shows that both silicon and Mg_2Si particles have been spheroidised by the heat treatment.

Table III shows the measured tensile properties and hardness of the alloy subjected to different heat treatments. A maximum yield strength of ~49 ksi is obtained when the alloy is aged at 170°C for 8 hours. This is a significant improvement over A390 alloy which has YS of about 45 ksi. It seems that ageing at higher temperatures is detrimental to the hardness of this alloy.

Table III. Room temperature tensile properties and hardness of Al-14Si-0.2Fe-3Mg-1MM-0.025Sr as a function of heat treatment.

Temper designation	Heat treatment	UTS (ksi)	YS (ksi)	% El.	Hardness (HRB)
As-cast	-	32.0	21.1	2.29	40
T4	1 day	40.7	30.4	2.46	60
T6	4 hours – 170°C	49.3	44.7	0.94	74
	8 hours – 170°C	50.3	48.9	0.67	78
T6	4 hours – 200°C	50.7	48.6	0.73	75
	8 hours – 200°C	46.7	43.5	0.79	70

Conclusions

A hypereutectic Al-Si-Mg-Fe alloy that exhibits a fine dispersion of Si and Mg_2Si particles is being developed. The tensile strength of this alloy is superior to that of A390 alloy, but its hardness is slightly lower than that of A390 alloy. Although the amount of misch metal used in the alloy (1wt%) may be considered somewhat high for commercial purposes, the amount required may be reduced considerably if higher cooling rates are employed, such as in die casting operations.

References

1   J.E. Hatch, Aluminum, Properties and Physical Metallurgy, ASM, Metals Park, OH, 1984, pp. 346 – 347.

2   S. Spigarelli, E. Evangelista, S. Cucchieri, Mater. Sci. Eng. A387–389 (2004) 702.

3   L. Lasa, J.M. Rodriguez-Ibabe, Mater. Sci. Eng. A363 (2003) 193.

4   J. Zhang, Z. Fan, Y.Q.Wang, B.L. Zhou, Mater Sci Eng. A281(2003) 104.

5   O. El Sebaie, A.M. Samuel, F.H. Samuel, H.W. Doty, Mater. Sci. Eng. A486 (2008) 241.

6   J. Zhang, Z. Fan, Y.Q. Wang, B.L. Zhou, Mater. Sci. Eng. A281 (2000) 104.

Aluminum Alloys: Fabrication, Characterization and Applications II
Edited by: Weimin Yin, Subodh K. Das, and Zhengdong Long
TMS (The Minerals, Metals & Materials Society), 2009

ELECTRON MICROSCOPY OF COMMERCIAL PURITY AL-2024 (Al-Mg-Cu) AFTER ACCUMULATIVE ROLL-BONDING

A.K. Kulovits[1], B. Webler[2], A.R. Deshpande[3] and J.M.K. Wiezorek[1]

[1] Department of Mechanical Engineering and Materials Science
University of Pittsburgh, 648 Benedum Hall, Pittsburgh PA 15261, USA;
[2] Department of Materials Science and Engineering
Carnegie Mellon, Pittsburgh PA 15213, USA
[3] GE Industrial and Commercial Lighting, Willoughby, OH 44094;

Keywords: Accumulative roll-bonding, Nanocrystalline Al – 2024, TEM

Abstract

Al-2024 an age-hardening Al – Cu – Mg alloy has been severely plastically deformed using accumulative roll bonding (ARB). Partial bonding has been achieved successfully. Changes in the microstructure have been investigated using imaging, diffraction and analytical methods of electron microscopy. Two different regions, namely an ultra-fine grained (UFG) region that contains elongated grains and regions comprised of equiaxed nanocrystalline (NC) grains, have been observed in the ARB product. Transmission electron microscopy (TEM) diffraction studies provided evidence of dynamic solid-state reactions for the NC regions, while the UFG regions exhibited the same major phases as prior to ARB. We report that the application of high strains in conjunction with dynamic transformations in the microstructure allows for grain refinement beyond the UFG regime into the NC size regime for this commercial purity Al-alloy.

Introduction

Severe plastic deformation (SPD) has been extensively studied and applied to obtain ultra fine grain (UFG) Al alloys. Different SPD schemes, e.g. equal channel angular pressing (ECAP) [1], accumulated roll bonding (ARB) [2] and severe plastic torsion straining (SPTS) [3], have been successfully applied to obtain UFG Al alloys. These UFG Al alloys possess better mechanical properties than their conventional coarser grain-size equivalents and good damping properties [4]. For example, J. Mao et al. report an improved 0.2% proof stress and ultimate tensile strength (UTS) in Al 2024, a commercially important, age-hardening alloy based on an Al-matrix and alloying additions of Cu and Mg (figure 1), without loss of ductility after ECAP [5]. A grain size refinement to a submicron (UFG) average grain size has been achieved by this technique [5]. However, the sample geometry that can be used in ECAP is rather limited and the process is currently not suitable for industrial applications that desire sheet metal components (e.g. automotive industry). ECAP is also labor intensive and the process throughput is less as compared to rolling based processes [6].

In this study ARB of Al-2024 (commercial purity) has been conducted. A strain of 7.5 has been introduced in the Al alloy using ARB. The microstructure of the alloy after 10 passes in ARB has been investigated using transmission electron microscopy (TEM) imaging and diffraction. The change in hardness during ARB has also been reported. Indications for dynamic precipitation of metastable precipitates during the process of ARB have been observed. An attempt has been made to identify the precipitates using electron diffraction in the TEM and comparison to previously published reports on the lattice parameters and crystal structures of the stable and metastable precipitate phases observed in conventionally processed Al-Cu-Mg alloys.

2. Experimental

The Al-2024 samples had a composition of $Al_{93.4\pm1.2wt.\%}Mg_{2.4\pm0.3wt.\%}Cu_{4.2\pm1wt.\%}$ as determined using X-ray energy dispersive spectroscopy (XEDS) with a JEOL JEM 2000FX TEM. Comparative microstructural analysis of the as received state and the SPD state after ARB were performed using the same TEM microscope and operating parameters. Bright field TEM imaging has been used to investigate changes in the morphology of the alloy microstructures. Electron diffraction in the TEM has been used for phase identification. Between successive passes during the ARB process the touching surfaces of the sheets in the stacks were wire brushed. Two sheets of the alloy with a starting thickness of nominally ~ 1mm per sheet were stacked together and reduced in thickness by 50% per pass. The newly formed sheet was cut in half; the halves were wire brushed; stacked together and then reduced in thickness by 50% again. This process has been repeated 10 times. The final stack of ARB foils contained 1024 layers. The Vickers hardness has been measured before and after ARB processing.

3. Results and Discussion

The composition ranges for Cu and Mg reported for Al-2024 are indicated in the ternary phase diagram section of figure 1 [7]. By XEDS TEM analysis we determined that the present alloy exhibited a slightly elevated Mg content relative to the range of nominal alloying element composition specified for Al-2024. The ternary phase diagram section (figure 1) indicates that the composition of the current alloy falls into the two-phase field $\alpha + S$. The possible precipitation sequence for an alloy located in this phase field has been determined previously [7] and is also reported for convenience in figure 1. The as-received Al-2024 alloy was in the T3 temper condition (solutionized plus subsequent cold work). The microstructural investigation of the as-received alloy revealed an average grain size of 45 ± 5 μm. The bright field (BF) TEM image (left panel figure 2) shows the presence of large dispersoids. The BF TEM image in the right panel of figure 2 shows pinned dislocations. These morphological observations are consistent with the T3 condition documented in prior reports for Al-2024 [8]. Electron diffraction in the TEM (figure 3, left panel) shows the presence of S phase in the α-Al matrix. Natural aging might be responsible for formation of this otherwise unexpected phase in the T3 condition. After 10 ARB passes (true strain in the foil is 7.5) each individual foil in the ARB stack of 1024 layers had a thickness of approximately ~850nm. We observed cracking on the surface of the foil, which can be rationalized by uneven bonding between

64

adjacent layers in the ARB sample. Nevertheless, on a macro scale the ARB sample appeared to act as a monolithic mass and showed a significantly improved mechanical hardness, Vickers number $H_V \approx 225$, corresponding to about 724MPa of tensile strength, with respect to the as received alloy, $H_V \approx 138$, corresponding to about 414MPa of tensile strength. Investigations were also conducted using TEM to elucidate the evolution of grain morphology during ARB. Two different types of regions have been observed.

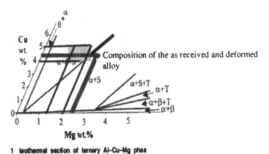

Possible precipitation sequence

(Supersaturated solid solution) SSS α → GP-zone → SÓ→ SÖ→ S (CuMgAl₂)(stable phase)

Figure 1: Isothermal section of the ternary Al-Cu-Zn phase diagram at 200°C (top) and possible precipitation sequences (bottom): blue box = stable phase window of 2024 Al at the room temperature; red lines = the actual composition of the present alloy

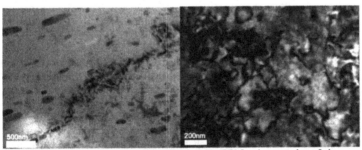

Figure 2: Multi-beam Bright Field (BF) TEM micrographs of the as received alloy: dispersoids (left) and pinned dislocations (right)

One part of the microstructure contained elongated grains, with sub-micron size (UFG), that are heavily defected (top row BF – images in Fig.-4). Dislocation mediated sub-grain formation, change in sub-grain boundary character from low- to high-angle grain boundary and subsequent grain refinement are processes typically expected to occur

65

during SPD [9], [10]. Electron diffraction shows regions that contain multiple grains of different orientations that are essentially α–Al (figure 3, bottom left panel). No evidence for precipitation differing from the state prior to SPD has been found in the diffraction data from this UFG microstructural constituent.

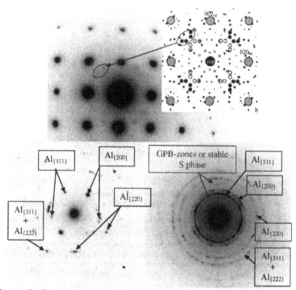

Figure 3: S̲elected A̲rea D̲iffraction (SAD) patterns: <100>-zone-axis of undeformed as received 2024 Al (from displayed area in Fig-2) and inset of simulated <100>-diffraction pattern with all possible S-phase variants (top row); SAD corresponding to the UFG region (Fig.-4) after ARB (bottom left); SAD of the nanocrystalline region (Fig-4) after ARB (bottom right)

The second type of microconstituent in the alloy microstructure after ARB contains nano-scaled grains (bottom row BF–images, Fig.-4). The average grain size in the nanocrystalline (NC) regions was determined as (28±10)nm with the maximum grain size not exceeding (75±10)nm. Selected area diffraction from the NC regions produced ring patterns (figure 3 bottom right panel). In comparison to the diffraction pattern of the UFG aggregate (figure 3 bottom left panel) taken from a region of equivalent area (figure 4 top and bottom right panels) a much larger number of diffraction maxima, i.e., of different orientations is present, consistent with the presence of a larger number of very small grains. The rings corresponding to $\{111\}_{Al}$ and $\{200\}_{Al}$ and other higher index planes have been identified as the prominent rings in the diffraction pattern (figure 3 right bottom panel). Additionally, diffuse diffraction rings are discernible (Figure 3). The lattice plane spacings associated with these diffuse diffraction rings are consistent with the presence of either GPB zones, metastable S", S' or stable S phase precipitates [11], [12], [13]. These latter diffuse intensity maxima must either be due to diffraction from entities that have emerged dynamically during the ARB deformation process or alternatively they could also stem from entities that were initially present in the microstructure and retained their crystal structures during ARB. Dark field TEM imaging to locate the precipitates was inconclusive. Additional HREM and analytical TEM is planned to further elucidate the evolution of precipitates at different locations in the morphologically inhomogeneous microstructure during the process of ARB.

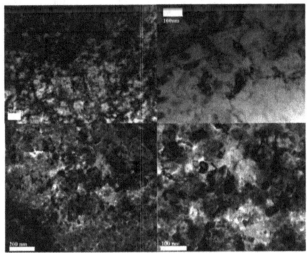

Figure 4: BF - TEM images after ARB: heavily defected area (top left); multiple elongated Ultra Fine Grains (top right) regions with Nano–Crystalline grains (~30-40nm) (bottom row)

SPD can lead to dissolution of precipitates, which in some instances can aid the precipitation process [14] or can lead to both, dissolution and subsequent re-precipitation. Thus it is not unreasonable to propose that the entities associated with the diffuse diffraction maxima we observed for the NC regions after ARB are metastable or stable precipitates or GPB zones that emerged during the cold deformation. Alternatively, they may have been inherited from the as received condition and underwent some significant morphological and scale changes due to the large amount of plastic deformation oft the ARB. Dynamic precipitation and recrystallization may have occurred preferentially in the interlayer bonding-interface area due to stress inhomogeneities arising from wire brushing induced surface roughness. Oxide layer fragments incorporation due to wire brushing may have further aided the pinning of otherwise mobile grain boundaries of dynamically recrystallized grains, enabling formation of NC microstructure during ARB. Differences in stress/strain evolution throughout layer interiors relative to the intralayer interface area during ARB provides a rationalization of the observation of two microconstituents that differ morphologically and in grain scale. A grain size refinement to the NC regime, using ARB for Al-2024 has been reported in here for the first time.

The increase in Vickers hardness associated with the ARB process corresponds to tensile strength increase from about 420MPa in the undeformed state to 720MPa after 10-

pass ARB. The hardness of the undeformed state is on the order of previously reported ultimate tensile strength (UTS) values for solutionized Al-2024 without artificial aging [8]. The hardness of the ARB state exhibits higher UTS values than those reported for Al-2024 after ECAP and subsequent naturally aging [8], but are lower than values reported after SPTS [15]. ECAP processed Al-2024 had a grain size slightly larger than the UFG grain size of the first microconstituent after ARB reported here (top row Fig-4.). The average grain size in SPTS processed Al 2024 was ~70nm. Both microconstituents are expected to contribute to the Vickers hardness we report here. Hence the significant hardness increase after ARB of Al-20204 may be attributed to contributions of grain refinement and work hardening in the UFG microconstituent and from grain refinement and dynamic solid-state reactions in the NC microconstituent. Additional experiments are required to develop detailed understanding of the solid-state reactions that we suspect to be ultimately responsible for the formation of NC regions in the ARB Al-2024.

4. Summary

ARB of Al-2024 resulted in a grain size refinement from an average grain size of 45μm in the as received condition to 28nm after 10 ARB passes in some regions of the microstructure. We observed regions (microconstituents) with a UFG (grain size <1μm) and a NC (grain size ≈28nm) grain size. The emergence of metastable precipitates during the ARB process has been proposed as a result of dynamic solid-state reactions. ARB increased the tensile strength by about 300MPa. We attributed the ARB induced 300MPa increase of the tensile strength to combined effects from grain refinement and work hardening in the ultra fine-grained microconstituent and from grain size refinement and dynamic phase transformation processes in the nanocrystalline microconstituent.

ACKNOWLEDGMENTS

The material presented in this article received partial support from the National Science Foundation under grant DMR-0094213, including part sponsorship of an undergraduate researcher (BW). Opinions, findings, and conclusions or recommendations expressed in this material are those of the authors.

References
1. Zhao Y H, Liao X Z, Jin Z, Valiev R Z, Zhu Y T. Acta Mater. (2004) 52, 4589.
2. Karlik M, Homola P, Slamova M. J Alloys and Compounds (2004) 378, 322.
3. Stolyarov V V, Shestakova L O, Zhu Y T, Valiev R Z. Nanostructured Materials (1999) 12, 923.
4. Koizumi Y, Ueyama M, Tsuji N, Minamino Y, Ota K. J Alloys and Compounds. (2003) 355, 47.
5. Mao J, Kang S B, Park J O. J Mater. Proc. Tech (2004) article in press.
6. Saito Y, Utsunomiya H, Tsuji N, Sakai T. Acta Mater. (1999) 47, 579.
7. S. C. Wang and M. J. Starink: International Materials Reviews 2005, VOL 50 NO 4 193
8. Kim W J, Chung C S, Ma D S, Hong S I, Kim H K. Scripta Mater. (2003) 49, 333.
9. Valiev R. Z., Alexandrov I.V. Ann. Chim. Sci. Mat. (2002) 27 (3), 3.
10. Lee S. H., Saito Y., Sakai T., Utsunomiya H. Mater. Sci. & Eng. A (2002) 325, 228
11. Ringer S P, Hono K, Polmear I J, Sakurai T. App. Surface. Sci. (1996) 94/95, 253.
12. Kovarik L, Gouma P I, Kisielowski C, Court S A, Mills M J. Acta Mater. (2004) 52, 2509.
13. Wang S C, Starink M J. Mater. Sci. & Eng. A (2004) A386, 156.
14. Embury J D, Deschamps A, Brechet Y. Scr. Mater (2003) 49, 927.
15. Krasilnikov N.A., Sharafutdiniv A. Mater. Sci. & Eng. A (2007) 463, 74.

Aluminum Alloys: Fabrication, Characterization and Applications II
Edited by: Weimin Yin, Subodh K. Das, and Zhengdong Long
TMS (The Minerals, Metals & Materials Society), 2009

ANNEALING BEHAVIOR OF AN HEAVILY DEFORMED ALUMINUM ALLOY IN 20 TESLA MAGNETIC FIELD

Samuel T. Adedokun[1], Olalere Ojo[2], O. Aluko[3]

[1]FAMU-FSU College of Engineering, Tallahassee, FL, USA
[2]Ladoke Akintola University of Technology, Ogbomoso, Oyo State, Nigeria
[3]Michigan Technological University, Houghton, MI, USA

Abstract

A plate of aluminum alloy 6061 was given 85% deformation by cold rolling. Samples from the rolled specimen were heat treated for different times and at different temperatures in a 20 Tesla resistive magnetic field. The effect of time and temperature on the heavily deformed specimens under 20 Tesla magnetic field was examined with the use of an Environmenetal Scanning Electron Microscopy equipped with an Orientation Imaging Microscopy (OIM) to study the changes in the grain size distribution and the grain boundary misorientation of the samples tested. The results indicate that the magnetic field of 20 Tesla increased the average grain size of the Aluminum alloy 6061 when compared with the aluminum alloy heat treated when the magnetic field was turned off. No effect on the grain boundary misorientation was noticed.

Introduction

This work presents the grain size distribution and the grain boundary misorientation data obtained after heat treating an heavily deformed aluminum alloy 6061 in a 20 Tesla resistive magnetic field and also after the magnetic field has been turned off. Aluminum alloy 6061 is a widely used material for structural and other applications. It properties after heat treatment is of importance to its applications in service. The objective of this study is to examine the effects of 20 Tesla magnetic field on the grain size distribution and grain boundary misorientation of aluminum alloy 6061 when heat treated at different temperatures and for different periods of time.

Experimental

A plate of aluminum alloy 6061 was heavily rolled down to 85% deformation from 13 mm to 2 mm thickness at room temperature. Samples from the rolled specimen were then heat treated at temperatures of 350°C, 400°C and 450°C for different time periods of 10, 20, 40 and 60 minutes in a 20 Tesla resistive magnetic field. The heat treatment was repeated after the magnetic field has been turned off at the same temperatures and for the same periods of time. Annealing of the samples was done with the rolling direction of each sample parallel to the direction of the magnetic field. This was done because it has been found that it produces maximum effect on the grain characteristics of the tested material.

The heat treated samples were thereafter mechanically polished to 0.05 micron surface finish and electropolished in a solution of 18 ml distilled water, 146 ml ethanol, 20 ml ethylene glycol monobutyl ether and 15.6 perchloric acid for about 30 seconds before being examined in an Electron Scanning Microscope equipped with Electron Back-Scattered Diffractometer (EBSD). The EBSD data was analyzed with a TSL© software to obtain the grain size distribution and grain boundary misorientation plots for the samples.

Results and Discussion

Results obtained after annealing for 10 and 40 minutes at the various temperatures are presented here. Figures 1 and 2 illustrate the grain size distribution for AA6061 samples annealed with/without magnetic fields at different temperatures for different periods of time as indicated in the figures.

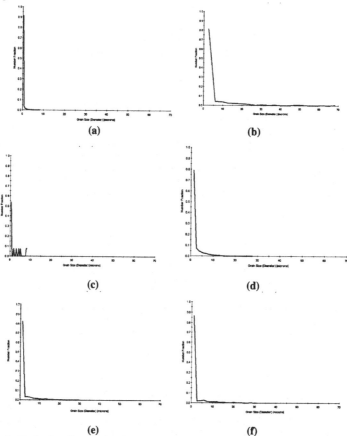

Figure 1: Grain size distribution for AA6061 samples annealed without magnetic field at (a) 350°C for 10 minutes, (b) 350°C for 40 minutes, (c) 400°C for 10 minutes, (d) 400°C for 40 minutes, (e) 450°C for 10 minutes and (f) 450°C for 40 minutes.

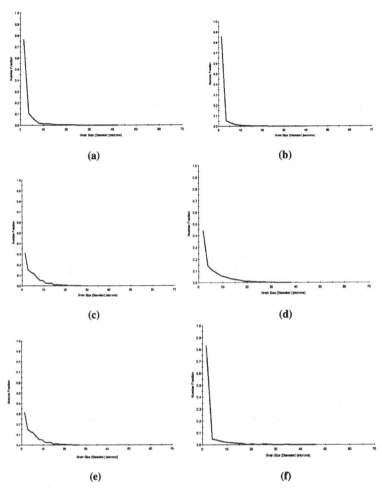

Figure 2: Grain size distribution for AA6061 samples annealed with 20 T magnetic field at (a) 350°C for 10 minutes, (b) 350°C for 40 minutes, (c) 400°C for 10 minutes, (d) 400°C for 40 minutes, (e) 450°C for 10 minutes and (f) 450°C for 40 minutes.

Grain misorientation angle distributions data are shown in Figures 3 and 4.

71

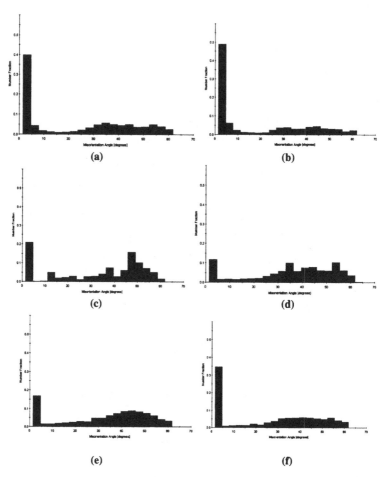

Figure 3: Misorientation angle distribution for AA6061 samples annealed without magnetic field at (a) 350°C for 10 minutes, (b) 350°C for 40 minutes, (c) 400°C for 10 minutes, (d) 400°C for 40 minutes, (e) 450°C for 10 minutes and (f) 450°C for 40 minutes.

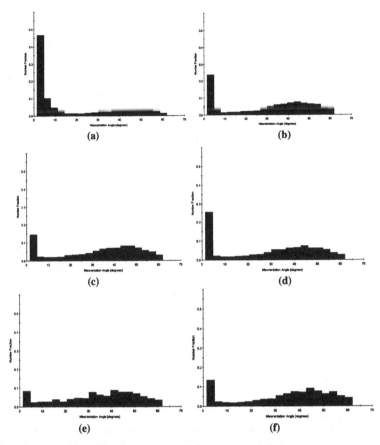

Figure 4: Misorientation angle distribution for AA6061 samples annealed with 20 Tesla magnetic field at (a) 350°C for 10 minutes, (b) 350°C for 40 minutes, (c) 400°C for 10 minutes, (d) 400°C for 40 minutes, (e) 450°C for 10 minutes and (f) 450°C for 40 minutes.

Grain size control can affect the performance and efficiency of a material in service. A look at diagrams in Figures 1 and 2 shows increase in grain size due to application of magnetic field for 10 minutes at 350°C. Similar trends occur for other tests carried out as shown in the plots in Figures 1 and 2 which led to increase in grain size. However, no conclusive deduction could be made for the grain misorientation distribution data.

73

Conclusions

It has been demonstrated that microstructure development during annealing of cold rolled aluminum alloy 6061 can be affected by a magnetic field. Magnetic annealing generally caused slight increase in the grain size of the aluminum alloy in 20 T resistive magnet. The major reason might not have been unconnected to magnetically induced interaction of dislocations with point defects which may lead to changes in the dislocation mobility depending on the test environment. It is known that a magnetic field can induce the motion of dislocations without any external mechanical stress [Al'shits et al, 1987].

Magnetic annealing still remains a difficult area with respect to understanding what happens to the processed material properties. Different authors have come up with different experimental results and propositions with various arguments to support their findings. More work need to be done in this area of research.

Acknowledgements

The author would like to thank Dr Reginald Perry (Associate Dean for Academics) at the FAMU-FSU College of Engineering, Tallahassee, Florida USA for providing funding through the Minority Doctoral Engineering Fellowship funded by Title III of the US Department of Education. Also, thanks to Dr Hamid Garmestani of the Materials Science and Engineering Department, Georgia Institute of Technology, Atlanta, Georgia and Dr Justin Schwartz of the Mechanical Engineering Department, Florida State University, Tallahassee, Florida for their academic advice.

References

1. S. T. Adedokun, J. Schwartz and H. Garmestani (2007) "Texture variation in an heat treated heavily deformed aluminum alloy 6061" Materials Science & Technology 2007 Conference Proceedings, pages 297 – 302.
2. Al'shits et al (1987) "Motion of dislocations in NaCl crystals under the action of a static magnetic field" *Sov. Phys. Solid State* **29**(2), February 1987.
3. S. Bhaumik, X. Molodova, D.A. Molodov and G. Gottstein (2006) "Magnetically enhanced recrystallization in an aluminum alloy" Scripta Materialia 55, pages 995-998.
4. Chen (1998) "Relationship between texture and magnetic properties of non-oriented Si steel" The Third Pacific Rim International Conference on Advanced Materials and Processing (PRICM 3), pp223-228.
5. D.A. Molodov, S. Bhaumik, X. Molodova and G. Gottstein (2006) 'Annealing behavior of cold rolled aluminum alloy in a high magnetic field' Scripta Materialia 54, pages 2161-2164.
6. A. D. Sheikh-Ali, D. A. Molodov and H. Garmestani (2002) 'Magnetically induced texture development in zinc alloy sheet' Scripta Materialia 46, pages 857-862.

EFFECTS OF ULTRASONIC TREATMENT ON MICROSTRUCTURES OF HYPEREUTECTIC AL-23%SI ALLOY

Haikuo Feng[1], Sirong Yu[1], Yinglong Li[2], Liyan Gong[2]

[1]Key Laboratory of Automobile Materials (Jilin University), Ministry of Education, and College of Materials Science and Engineering, Jilin University;
No.5988 Renmin Street, Changchun 130025, P R China;
[2]School of Materials and Metallurgy, Northeastern University, Shenyang 110004 China

Keywords: Ultrasonic treatment, Hypereutectic Al-Si; Primary Si

Abstract

Microstructures of hypereutectic Al-23wt%Si alloys were gained with and without ultrasonic treatment in a novel horn crucible designed specially for this experiment. Evolution, morphology and distribution of microstructure of alloys were investigated. The results show that the size of primary Si particulates decreased from 500μm to 180μm and the morphology of α-Al phase had been changed from dendritic crystal to equiaxial crystal under the ultrasonic treatment. However eutectic phase was coarser than that without ultrasonic treatment. The mechanisms of alloys treated by ultrasonic were discussed.

1. Introduction

Hypereutectic Al-Si cast alloys are especially suitable for tribological parts owing to their excellent wear resistance provided by the presence of silicon [1-3]. However, the machinability and ductility of hypereutectic Al-Si alloys are poor. The mechanical properties of Al-Si alloys depended mainly on the size, shape and distribution of Si, and not on the content of silicon in the alloys. It is effective to change the morphology and size of Si in order to decrease the weakening effect of Si on the matrix and improve the properties of Al-Si alloys. The Al-Si alloy will excellent mechanical properties as long as silicon phases, either the eutectic Si or primary Si, are fine and distribute uniformly in matrix [2,3].

Controlling the microstructure of castings is considered as one of the main challenges faced by today's foundry industry. Mechanical modification, including electromagnetic stirring or mechanical vibration to induce forced convection in the melt, has been developed and applied successfully in the industry [4]. It remains desirable to find a method, which can provide the benefits of modification combined with good castability, simple operation and persistence of the modified structure on alloys [2]. Ultrasonic treatment as a new technology is being used widely for improving the solidification structure of Al-Si alloys. The injection of ultrasonic energy into molten alloys can bring about some nonlinear effects, such as cavitation, acoustic stream, which can be used to refine microstructures, reduce segregation, and degas [5,6].

The goal of this work is to investigate the influence of ultrasonic treatment on the hypereutectic Al-Si alloy with a novel ultrasonic treatment setup and to develop a new simple and economical method to modify the hypereutectic Al-Si alloys.

2. Experimental

2.1 Materials

Commercial hypereutectic Al-Si alloy ingots were used as raw material. Its chemical composition was listed in Table 1. Alloy ingots were melted using an intelligently numerical controlled electric resistance furnace.

Table 1 Chemical composition of Al-Si alloy (wt.%)

Si	Mn	Cu	Ti	Fe	Zn	Mg	Al
23	0.01	0.08	0.02	0.5	0.005	0.101	balance

2.2 Ultrasonic treatment equipment

The schematic diagram of the apparatus is shown in Figure 1[7]. The ultrasonic treatment equipment consists of the temperature and power control systems. The treating temperature was monitored with a temperature sensor.

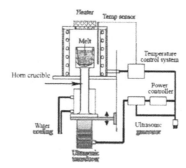

Figure 1. Schematic diagram of ultrasonic treatment equipment

Horn was one of the important parts of ultrasonic treatment equipment. The ordinary horn included four styles, such as stepped horn, linear horn, exponential horn, and catenoidal horn [8]. A special horn (in Fig.1) was designed and fabricated, and there was a crucible at its end. So, the special horn was named horn crucible. The length and cross-section area of the horn crucible were determined according to the theory that the stress passes continuously through the length of the horn [8,9]. The horn crucible was vertically bolted with a 20 kHz transducer.

The ultrasonic emitted from the transducer and passed through the acoustic horn (horn crucible) was propagated directly into the melt. In this case, the alloy melt became a part of acoustic horn, so the action of acoustic energy on the melt was raised remarkably.

2.3 Experimental procedures of ultrasonic treatment

Al-Si alloy of 200g was heated to 800°C and then held at this temperature for 30min. The melt was stirred slightly and poured into the horn crucible whose temperature was 680°C,and then was treated for 5min and10min without and with ultrasonic. The ultrasonic power is 50W and the amplitude is 4μm. After that, the horn crucible was moved out the temperature control system and was immediately quenched with the water of 25°C.

76

2.4 Observations of structure and properties

Microstructures of the samples were investigated using Olympus GX51 optical microscope and the sizes of various phases were measured statistically by quantitative metallography analysis method.

2. Results and discussion

Fig.2 shows the microstructures of Al-23%Si alloy without and with ultrasonic treatment. The distribution of primary Si particles in hypereutectic Al-Si alloy not treated by ultrasonic was homogeneous, and no aggregation was found in large scale over the section of the sample (Fig.2(a)). The shapes of primary Si phases included mostly coarse flower crystal, polygon, rod and blocky crystal. The flower crystal was commonly composed of 5-6 club shaped crystal. The edges and corners of primary Si phases were clear. The largest size of primary Si phases was up to 500μm (Fig.3). After hypereutectic Al-Si alloy was treated by the ultrasonic in 5 min, the size of primary Si was smaller than that without ultrasonic (Fig.2(b)). But in the ultrasonic processing time of 10min, most of the primary Si particles distributed homogeneously in small scale over the section of the sample (Fig.2(c)). As compared with the microstructures without ultrasonic treatment, the large irregularly shaped primary Si phases disappeared, and at the same time the size of primary Si phases obviously decreased to 180μm. The morphology of six clubbed flower was unconspicuous and the edges of primary Si phases change to circle smooth. The reasons that the microstructures were changed include several aspects. Firstly, the compress and relaxation of high frequency ultrasonic have effect on the melt, so the edge of particles in the melt would be scoured and form a circle surface. Secondly, transient cavitation could produce an impact strong enough to break up the clustered fine particles and disperse them more uniformly in melt. Third, the strong impact coupled with locally instantaneously high temperature in a very short time could also remelt the primary Si phases so their edges would be circle of edges [10].

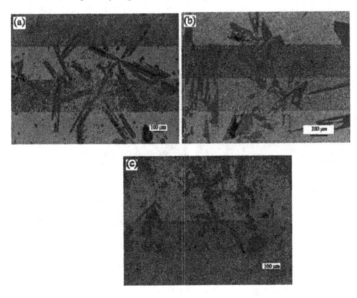

Figure 2.Effect of ultrasonic processing time on morphologies of primary Si in Al-23%Si
alloy (a)without ultrasonic treatment and ultrasonic treatment (b)5min, (c)10min

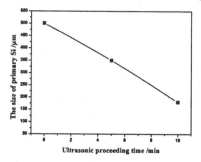

Figure 3.Effects on the size of primary Si with different ultrasonic processing time

Under the condition of non-equilibrium solidification, α-Al phase can also produce in
hypereutectic Al-Si alloys. It can be seen that the α-Al phase was developed obviously into
dendritic crystal without ultrasonic treatment and the length of its primary arm was even up to
120μm (Fig.4(a)). After the alloy was treated by ultrasonic, the morphology of the primary α-Al
dendritic crystal was changed to equiaxial crystal (Fig.4(c)), and the size of the primary α-Al
phase reduced obviously from 120μm to about 20μm. The modification of α-Al phase was
mainly due to the action of the caviation and acoustic stream of ultrasonic. Transient caviation
produced strong impact and broke up the α-Al dendritic crystal, and then acoustic stream
dispersed them uniformly to the melt. So the growth of α-Al dendritic crystal was restrained and
resulted in the formation of α-Al equiaxial crystal.

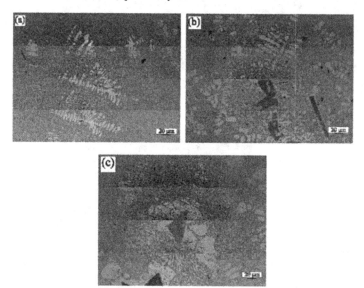

Figure 4.Effect of ultrasonic processing time on morphologies of α-Al in Al-23%Si alloy (a)without ultrasonic treatment and ultrasonic treatment (b)5min, (c)10min

The shape of eutectic silicon in hypereutectic Al-Si alloys without ultrasonic treatment is observed to be flaky. The lamellar spacing of the eutectic structure is about 2.2μm(Fig.5(a)). The ultrasonic treatment is effective in restraining the emanation of primary Si phase, and helping the silicon to dissolve into the melt and to form the irregular eutectic, as well as aiding the breakage of dendritic crystal [4].When the ultrasonic transmitted into the melt, the vibration of high frequency was induced in the melt and which would influence and adjust the solution of alloys. The growth of primary Si in anisotropy was restrained and enforced to change the nucleation and growth by the ultrasonic treatment. The effect of ultrasonic treatment could have tended to weaken the covalent bond of Si and Si, so that the primary Si was restrained for priority to precipitation. The eutectic lamellar spacing increased slightly in size as compared with that of the non-ultrasonic treating sample from 2.2μm to 2.8μm of that with ultrasonic treatment in processing time of 10min. The stirring melt under the expanding during the negative pressure cycle and collapsing during the positive pressure cycle increased the diffusion of solute, at same time the content of silicon was high. The enrichment of solute was large in the foreland interface of solidification, which made the degree of supercooling increase and led the temperature of growth decrease. This is the reason the eutectic lamellar spacing increased with ultrasonic treatment. The ultrasonic treatment affected the microconstituents in the eutectic phase very significantly, and therefore it was considered to be the most important element determining the hardness and the abrasive wear resistance [11].

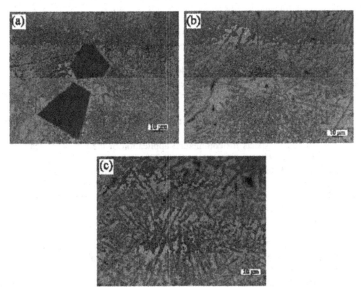

Figure 5.Effect of ultrasonic processing time on morphologies of eutectic in Al-23%Si alloy (a)without ultrasonic treatment and ultrasonic treatment (b)5min, (c)10min

Some gas was found in the alloy untreated by ultrasonic (Fig. 2(a)). This indicates that the gas bubbles could not floated out the alloy melt in the course of cooling and solidification.

However, the gas holes were not found in the alloy treated with ultrasonic (Fig. 2(b)and 2(c)). This is because the acoustic flow induced by ultrasonic helped the coagulation and float of hydrogen bubbles. The action of degassing of ultrasonic was also found by other researchers [6].

3. Conclusions

Microstructures of hypereutectic Al-23%Si alloy were fined and homogenized by ultrasonic treatment. The size of primary Si particulates decreased from 500μm to 180μm. Ultrasonic treatment successfully broke the dendritic structure of α-Al phase into equiaxial crystal, and the primary arm size was decreased from 120μm to 15μm. The size of eutectic was increased slightly from 2.2μm to 2.8μm

Acknowledgements

This work was supported by The Specialized Scientific Research Foundation for Doctor Subject in Colleges and Universities by the Ministry of Education of China (Grant no. 20030183019), Program for New Century Excellent Talents in University, and 985 project of Jilin University of China.

References

[1] M.M. Haque, M.A. Maleque, and J. Mater, "Effect of process variables on structure and properties of aluminium–silicon piston alloy", *Journal of Materials Processing Technology*, 77 (1998), 122-128.

[2] S. Tomida, et al., "Improvement in wear resistance of hyper-eutectic Al-Si cast alloy by laser surface remelting", *Surface and Coatings Technology*, 169-170(2),(2003),468-471.

[3] L. Lasa, J.M. and Rodriguez-Ibabe, "Effect of composition and processing route on the wear behaviour of Al-Si alloys", *Scripa Materialia*, 46 (2002), 477-481.

[4] R.Y. Wang, W.H. Lu, and L.M. Hogan, "Self-modification in direct electrolytic Al–Si alloys (DEASA) and its structural inheritance", *Materials Science and Engineering A*, 348 (2003), 289-298.

[5] X. Jian, et al., "Effect of power ultrasound on solidification of aluminum A356 alloy", *Materials Letter*,59 (2-3) (2005),190-193.

[6] G.I. Eskin, "Broad prospects for commercial application of the ultrasonic (cavitation) melt treatment of light alloys", *Ultrason Sonochem*, 8 (3) 2001,319-325.

[7] H.K. Feng, et al., "Effect of ultrasonic treatment on microstructures of hypereutectic Al–Si alloy", *Journal of materials processing technology,* doi:10.1016/j.jmatprotec.2007.12.121

[8] Amit Lal, and R.M. White, "Silicon microfabricated horns for power ultrasonics", *Sensors and Actuators A*, 54 (1996), 542-546.

[9] S.Y. Lin, "Load characteristics of high power sandwich piezoelectric ultrasonic transducers", *ultrasonics*,43(5)(2005), 365-373.

[10] Y. Yang, J. Lan, and X.C. Li, "Study on bulk aluminum matrix nano-composite fabricated by ultrasonic dispersion of nano-sized SiC particles in molten aluminum alloy", *Materials Science and Engineering A*, 380(1-2)(2004), 378-383.

[11] D.K. Dwivedi, "Microstructure and abrasive wear behaviour of iron base hardfacing", *Materials Science Technology*, 20(2004), 1326-1330.

Aluminum Alloys: Fabrication, Characterization and Applications II
Edited by: Weimin Yin, Subodh K. Das, and Zhengdong Long
TMS (The Minerals, Metals & Materials Society), 2009

SOLIDIFICATION AND PROCESSING OF ALUMINUM BASED IMMISCIBLE ALLOYS

H.R. Kotadia[1], J.B. Patel[1], Z. Fan[1], E. Doernberg[2], and R. Schmid-Fetzer[2]

[1] BCAST (Brunel Centre for Advanced Solidification Technology),
Brunel University, Uxbridge, UB8 3PH, UK
[2] Clausthal University of Technology, Institute of Metallurgy,
Robert-Koch-Str. 42, D-38678 Clausthal-Zellerfeld, Germany

Keywords: Immiscible alloys, Solidification, Microstructure, Intensive shearing

Abstract

The Al-Sn and Al-Pb based immiscible alloys have significant potential for bearing applications. However, the mixing and understanding of solidification process for immiscible alloys have been long standing challenges for their development. This paper presents solidification and microstructural evolution of the Al-Sn-Cu alloys and also describes the mechanism of effective mixing by the intensive shearing. The experimental work was also focused on analyzing the effects of shear rate, temperature and time on Sn droplets size and their distribution. Results have been compared with earlier study on Al-Si-Pb alloys. Experimental results suggest that the intensive shearing process produces homogeneous and finely dispersed Sn and Pb droplets.

Introduction

The solidification studies of immiscible alloy systems such as Al-Bi, Al-Sn, Al-Pb, Al-Si-Pb, Al-Pb-Si etc. is important from scientific and technical point of view [1-3]. The Al-Sn and Al-Pb based alloys have been commonly accepted for having excellent tribological and mechanical properties. These kinds of alloy system are suitable for engineering applications, particularly self lubrication bearing materials [1, 2]. Owing to the lower solubility, the parent liquid is decomposed into two distinct immiscible liquid phases when it passes through the immiscibility gap [1-3], and then followed by severe segregation due to the large density difference between two different density liquid phases [1,2]. In Al-Sn and Al-Pb alloy systems, phase separation occur when the Sn and Pb content are higher than 0.09 wt.% and 0.2 wt.%, respectively. To overcome segregation problem in immiscible alloy many methods have been proposed, such as stir casting, ultrasonic, rheocasting and rapid solidification. Recently, Fan et. al [4] developed a melt conditioning advanced shearing technology (MCAST) device to create a fine and homogeneous liquid dispersion within the miscibility gap and then the viscous force offered by semi-solid slurry to counterbalance the gravity force and the Marangoni effect [4,6].

In the present study, the immiscible Al-Sn-Cu alloys were successfully synthesized within the semi-solid region using the well developed MCAST device and results are compared with earlier study on Al-Si-Pb alloys system. It is observed that the final microstructures of alloys are strongly influenced by the viscosity of the system, shear forces, turbulence and cooling rate.

Experimental procedure

The (90-x)Al-xSn-10Cu immiscible alloys for x = 20, 30 and 45 were prepared from commercial pure aluminium with appropriate addition of 99.99 wt.% pure Sn and Cu and Al-Si-xPb alloys for x = 3.8, 7.2 and 17.2, were prepared from A357 alloy with appropriate addition of 99.97 wt% pure Pb [6]. All compositions in this paper are given in wt%. The melt was prepared in a graphite clay crucible in electric resistance furnace. The furnace temperature was gradually increased and held 200 °C above critical temperature (T_c) for 2 hours to homogenize the melt.

The MCAST device used in this work for intensive shearing is combined with high pressure die casting (HPDC) machine (DCC280, LK® Machinery, Hong Kong). The combination of MCAST and HPDC is called MC-HPDC. The detailed explanation about MCAST has been described elsewhere [7, 8]. The Al-Sn-Cu and Al-Si-Pb alloys melt were poured into the MCAST device at 650 °C and 620 °C, respectively. The pouring temperature was well above the T_c to avoid phase separation of L′ and L″ (L′ is Al-rich liquid and L″ is Sn-rich or Pb-rich liquid) before the shearing commenced and then multi-phase mixture was sheared at desired speed, time and processing temperature (T_p). For microstructural comparison purpose melt was directly transferred to the HPDC machine without shearing, which is referred to as conventional HPDC process.

To investigate microstructural features with optical microscope (OM) the samples were mounted and ground using standard metallographic polishing techniques. In the process of microstructural characterization, the equivalent diameter (d) and shape factor (F) were calculated by $d = \sqrt{4A/\pi}$ and $F = 4\pi A / P^2$; where, A is the total area and P is the peripheral length of the particles. When F is equal to 1, it represents a perfect spherical particle.

Results

HPDC

Figure 2 shows the OM image of Al-45Sn-10Cu alloy produced by conventional HPDC process. Segregation of the Sn droplets (dark grey in contrast) can be seen at the centre of the tensile specimen. Due to presence of temperature gradient during solidification Sn droplets migrate from low temperature region to the high temperature region [9].

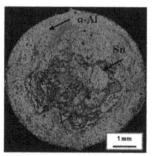

Figure 2. The cross section view of Al-45Sn-10Cu alloy
tensile specimen produced with conventional HPDC.

82

MC-HPDC

Al-Sn-10Cu alloys. Figure 3 shows the microstructures of Al-Sn-Cu samples produced with MC-HPDC. The Sn droplets (dark grey in contrast) in all samples are dispersed uniformly in Al matrix. A good distribution and fine size of Sn droplets achieved at optimum processing parameters. As the wt.% of Sn increases the average Sn droplets size increases from 4 µm to 22 µm with almost constant shape factor (Figure 4(a)). No significant segregation has been observed throughout cross section of the tensile specimen as shown in Figure 3. Figure 4(b) reveals that α-Al particles are also spherical in shape. The size of the α-Al particles vary between 40 µm to 50 µm with different Sn concentrations. The microstructures produced after shearing with varied shearing time and intensity have been also characterised for their respective volume fraction of α-Al particles and Sn droplets [9].

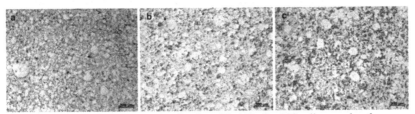

Figure 3. Optical micrographs of various (90-x)Al–xSn–10Cu alloys produced by the MC-HPDC process under optimal processing parameters (a) $x = 20$; $T_p = 580$ °C; shearing speed 800 rpm for 60 s (b) $x = 30$; $T_p = 580$ °C; shearing speed 800 rpm for 60 s(c) $x = 45$; $T_p = 535$ °C; shearing speed 800 rpm for 180 s.

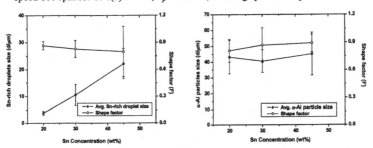

Figure 4. Effect of intensive shearing on (a) Sn droplets size and shape (b) α-Al particles size and shape as a function of Sn concentration.

Al-Si-Pb alloys. The resultant microstructures of Al-Si-Pb alloys are shown in Figure 5. The average size of the Pb droplets (black color in contrast) in Al-Si-3.8Pb alloy is 2.6 µm and the shape factor of the Pb droplets is 0.89. In Figure 6(a) by increasing Pb concentration from 3.8 wt.% to 17.2 wt.% the droplet size increases from 2.6 µm to 14 µm, while shape factor has decreased from 0.89 to 0.82. Similar to the Al-Sn-Cu alloys system, there has been no significant segregation found. In addition, primary α-Al particle are observed to distribute homogeneously and finely throughout the sample along with the uniform and well distributed Pb droplets. The size and shape of the α-Al particles have not changed much with increasing the wt.% of Pb in these alloys (Figure 6(b)).

Figure 5. Optical micrographs of various Al–Si–xPb alloys produced by the MC-HPDC process under optimal processing parameters (a) $x = 3.8$; $T_p = 605$ °C; shearing speed 500 rpm for 120 s (b) $x = 7.2$; $T_p = 605$ °C; shearing speed 500 rpm for 120 s (c) $x = 17.2$; $T_p = 595$ °C; shearing speed 500 rpm for 120 s [6].

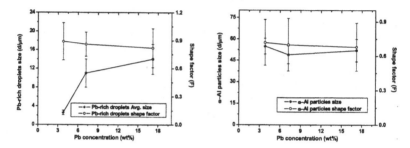

Figure 6. Effect of intensive shearing on (a) Pb droplets size and shape (b) α-Al particles size and shape as function of Pb concentration.

Discussion

In HPDC produced alloys, the higher density Sn droplets were accumulated at the central area of tensile specimen. The volume fraction of droplets and their size was increased from the mould wall to centre position of mould. This occurs because nucleation starts at the surface of mould and proceeds inward, but due to the migrating nature of the droplets from a low temperature region to high temperature region, the segregation occurs at the centre of the specimen at place where liquid solidifies last [1,2,10,11], which is described by the well known Marangoni motion. The coalescence mechanism mainly depends on the size and volume fraction of the L′ droplets [1, 11]. The coalescence takes place by the transfer of matter in which larger droplets grow by absorbing smaller ones and some droplets collide with each other to from a single one by mutual loss of surface energy due to joining [11].

In the MC-HPDC process, when the liquid alloy is fed into the MCAST (above the T_c) and the melt cools quickly to the barrel temperature set by the control system, which is usually just below the monotectic temperature (T_m), where primary α-Al already start to precipitate. At the same time, the melt separates rapidly into two immiscible liquids through nucleation and growth of liquid droplets in miscibility gap. Under the intensive shear mixing action created by the twin

84

screws, the liquid droplets attain fine particle size, as a result of the dynamic equilibrium between two opposite processes, coagulation and breakup of liquid droplets. The stages of the process from the homogeneous liquid to fine and uniformly distributed Sn and Pb droplets in Al matrix is shown schematically in Figure 7.

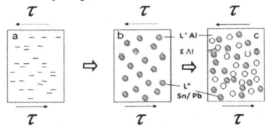

Figure 7. Schematic illustration of the rheomixing process for achiving a uniform distribution of soft phase in Al alloy matrix (a) homogenious liquid (above the T_c); (b) creation of the L″ droplets in L′; (c) rheomixing: formation of a primary α-Al solid phase (S) in L′ through a monotectic reaction.

The final size of liquid droplets will be dictated by the intensity of shear mixing action and the thermo-physical properties of the system, such as viscosity, interfacial tension, etc. When the melt reaches a temperature below the T_m, a solid phase will form from one of the liquid phases through the monotectic reaction [5]. It is well known that the viscosity of the semi-solid slurry increase exponentially with the volume fraction of the solid phase and decrease dramatically with increasing shearing rate and shearing time. By careful selection of the processing temperature, the viscous force was kept high enough to counterbalance the gravity force. Consequently, the alloy system is stabilized for the final solidification of the remaining liquid, normally by a eutectic reaction at a lower temperature. Therefore, viscosity helps to inhibit agglomeration or to slow down diffusion of the Sn and Pb droplets. The effect of viscosity of the semi-solid slurry on Stokes motion (U_s) and Marangoni motion (U_m) is given by:

$$U_s = \frac{2g\Delta\rho(\eta+\eta')}{3\eta(2\eta+3\eta')}r^2,$$ (1)

$$U_m = \frac{2\left|\frac{dT}{dx}\right|\left|\frac{d\sigma}{dT}\right|\kappa}{(2\eta+3\eta')(2\kappa+\kappa')}r,$$ (2)

Where, $\Delta\rho$ is the density difference between the two liquids, g is gravitational acceleration, r is the size of the liquid droplet, κ and κ' are conductivity of liquid matrix and droplets respectively; η and η' are viscosities of the liquid matrix and droplets respectively. dT/dx is the temperature gradient and $d\sigma/dT$ is the variation of the interfacial energy between the two liquid phases with change in temperature. During intensive shearing, the melt temperature is extremely uniform throughout the entire volume of the liquid mixture. According to equation (2), $U_m = 0$, therefore segregation of L″ droplets are negligible during solidification of intensively sheared melt.

The initial size distribution of droplets is inhomogeneous in the melt conditioner. Refinement and dispersion of droplets occur at later stages when increased the time of shearing in Al-Sn-Cu alloy [9]. The observed decrease in droplet size with shear rate is related not only to the breakup process but also to the shear-induced coalescence. The coalescence can be accelerated by the same factors that favor the drop breakup, i.e. high shear rate and reduced viscosity ratio. Therefore, the minimum droplet size under given shear mixing conditions is a dynamic balance between two opposite processes, droplet breakup and coalescence.

Summary

1. The MC-HPDC process produces a uniform dispersion of Sn and Pb droplets in Al alloy matrix. The size of the α-Al primary phase is approximately 50 μm and the average size of the Sn and Pb droplets increases with Sn and Pb concentration.

2. The Sn and Pb metallic droplets can be broken up more easily in the viscous fluid under high shear rate conditions and can achieve more spherical shape in thick viscous turbulent flow. Increasing shear rate speed up the droplets breakup process and will also lead to the spherical and fine droplet formation.

Acknowledgment

This work has been funded by the EPSRC and DTI. The authors would like to thank Dr. N. Hari Babu for their advice and encouragement.

References

1. L. Ratke and S. Diefenbach, "Liquid immiscible alloys," Mater. Sci. and Eng., R15 (1995), 263-347.
2. L. Ratke: Immiscible liquid metals and organic (The proceeding of immiscible alloys, Physikzentrum, Bad Honnef 1992).
3. J.Z. Zhao and L. Ratke, "A model describing the microstructure evolution during a cooling of immiscible alloys in miscibility gap," Scripta Materialia, 50(2004), 543-546.
4. Z. Fan, S. Ji, and J. Zhang, "Processing of immiscible metallic alloys by rheomixing process," Mater. Sci. and Tech., 17 (2001), 837-841
5. D. Mirkovi´c, J. Gr¨obner, and R. Schmid-Fetzer, "Liquid demixing and microstructure formation in ternary Al-Sn-Cu alloys," Mater. Sci. and Eng., 1-2(2008), 456-467.
6. X. Fang and Z. Fan, "Rheo-diecasting of Al–Si–Pb immiscible alloys," Scripta Materialia, 54 (2006), 789–793.
7. Z. Fan, M.J. Bevis, and S. Ji, PCT Patent, WO 01/21343 A1; 1999.
8. H. Tang, L.C. Wrobel, and Z. Fan, "Hydrodynamic analysis of binary immiscible metallurgical flow in a novel mixing process: rheomixing," Appl. Phys. A, 81A (2005), 549-559.
9. H.R. Kotadia et al., "Processing of Al-45Sn-10Cu based immiscible alloy by a rheomixing process," Solid State Phenomena, 141-143 (2008), 529-534.
10. A. Munitz and R. Abbaschian, "Microstructure of Cu-Co alloy solidified at various supercooling," Metall. Meter. Trans. A, 27A (1996), 4049-4059.
11. M.B. Robinson et al., "Undercooling, liquid separation and solidification of Co-Cu alloys," J. of Mater. Sci., 34(1999), 3747-3753.

Aluminum Alloys:
Fabrication, Characterization and Applications II

Formability and Texture

Session Chair

Subodh K. Das

Aluminum Alloys: Fabrication, Characterization and Applications II
Edited by: Weimin Yin, Subodh K. Das, and Zhengdong Long
TMS (The Minerals, Metals & Materials Society), 2009

BENDING PERFORMANCE OF Al-Mg-Si ALLOY AFTER INTERRUPTED AND DELAYED QUENCH

C. Bezençon[1], J.F. Despois[1], J. Timm[1], A.K. Gupta[2], C. Bassi[1]

[1]Novelis Innovation Center Sierre (NICS); CH-3960 Sierre, Switzerland
[2]Novelis Global Technological Center (NGTC); Kingston, Canada

Keywords: Bending, Automotive Alloys, Water Quench, Precipitation

Abstract

An improvement in the hemming performance of 6xxx alloys is a key requirement for automotive panel applications. The tensile and forming properties of these alloys can be improved by optimizing the alloy composition and the thermo-mechanical processing. In this paper, the influence of cooling conditions after solutionizing on the bending performance of commercial Al-Mg-Si alloys is reported. Quenching is performed through a water cooling system, which has the ability to control the temperature at which water quench starts and ends. The bending performance of material subjected to a wide range of quenching conditions was assessed via three-point bending tests and microstructural examination. It is shown that bendability is highly dependent on the quenching procedure within a critical temperature range and that bending performance correlates with grain boundary precipitation.

1. Introduction

Aluminium alloys are now increasingly replacing steel hang on panels such as motorhoods, fenders, roof and decklid by most of the automotive manufactures in Europe. This growth of aluminium applications has been achieved through the use of 6xxx alloys series, containing Si and Mg as main alloying elements. A variety of alloys like AA6016, AA6181A and AA6111 are now in use as they offer a good combination of high forming performance in T4 supply temper and excellent strength when the formed panel is exposed to the paint cure cycle.

In coming years, aluminium sheet alloys are expected to be used for more complex designs and new applications such as doors and roofs. Body designs are expected to be more aggressive, requiring better forming and sharp flat hemming aluminium sheet. Most of the alloys used for automotive applications will need improvement in hemming characteristics for such applications.

The effects of composition on precipitation and strengthening of 6xxx alloys is generally well understood and have been published in the literature [1-4]. On the otherhand, there have been only a few limited studies focusing on improving and understanding the bending performance of this alloy family [5-7]. Recently, Himura et al. [5] have reported that the quenching rate from 8 to 44 K/s and the final quench temperature range within 100-200 °C plays an important role in improving the bendability of the 6xxx alloys. The aim of the present study is to expand the scope with faster cooling rate (200 to 900 K/s), with a delayed start of water quench and various 6xxx alloys.

2. Experimental procedure

Figure 1 shows a schematic of the solution heat treatment and quenching set used in this study. In all experiments, a 1.0mm thick A4 size AA6016 sheet is introduced in a pre-heated furnace (590 °C in this study) to solutionized at the desired temperature and cooled down at a controlled rate adjusted through different modes: (i) natural cooling in ambient air, (ii) forced air cooling with a fan, (iii) water cooling or (iv) mixed mode air-water. The specimens were instrumented with a thermocouple linked to a data acquisition system sampling at 10 KHz with an acquisition rate of 100 Hz. Figure 2a shows a thermal profile recorded in a typical solutionizing and quenching cycle. Figure 2b shows the cooling rate agains temperature depicting start (Qstart) and end (Qend) temperatures of water cooling. Water quench rate QR is defined as the average slope of the t-T curve between Qstart and Qend. All samples were forced air cool beyond the Qend temperature.

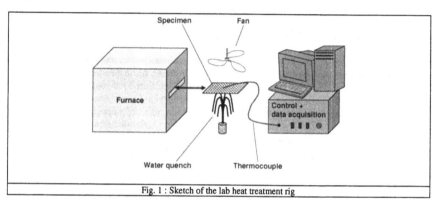

Fig. 1 : Sketch of the lab heat treatment rig

The cooling rate obtained with the quench apparatus is a function of the temperature. Fig. 2b shows three distinct cooling rate regimes: (i) Pulsed air regime : before Qstart where cooling rate ranges between 10 and 30 K/s, (ii) Vapour regime : between Qstart and 300°C were cooling rate is between 100 and 300 K/s and (iii) Water regime : below 300°C where cooling rate is > 500 K/s. The cooling rate between 450 and 300°C is fairly stable because of the formation of a vapour barrier between water and the metal, as a consequence of the Leidenfrost effect. Below a critical temperature (below 300 °C in this case), the vapour film breaks up and direct water contact to aluminium happens. The water cooling is then difficult to control at a fixed rate. Therefore an ending temperature of quench between 200 and 300°C cannot be obtained with the set-up shown in Fig.1. The quench profile shown in Fig. 2b was highly reproducible and exhibited the transition between each regime at similar temperature (around 300 °C, see figure 4 for example).

The hemming performance of the specimens was determined from the three point bending tests [8]. Prior to bending all specimens were uni-axially stretched in tension by 10% to simulate forming process. During the test, a sample of 30 x 50 mm^2 was forced between rolls of 30 mm in diameter by a punch with a tip radius of 0.2 mm. The bending angle reported in this paper represents the angle of the specimens when a 15N force drop during the test is noticed.

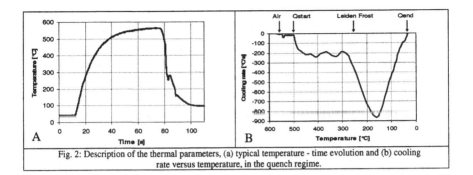

Fig. 2: Description of the thermal parameters, (a) typical temperature - time evolution and (b) cooling rate versus temperature, in the quench regime.

As a decrease in the force is generally linked to specimen failure (crack), the interpretation of the test is done so that low angle of bending means good bending behaviour and high value of angle means poor bending performance. Each point shown on the graphs (Fig. 5) is the mean value of 6 bending angles measured on specimens from the same batch. The dispersion for one batch is in the order of $\pm 4°$.

The failed specimens were polished and etched with dilute 0.25%-HF and examined in a LEO 1500 FEG Scanning Electron Microscopy (SEM), in secondary electrons mode.

3. Results and discussion

All specimens where solutionized at 567°C for 80±10s and quenched via 5 different schemes shown in fig. 3. The first two quench sequence (a) and (b) are reference conditions, showing the limits of properties whereas (c), (d), and (e) schemes using mixed air/water were done to define the critical temperature affecting bending.

Table 1 lists the cooling conditions and the bend data for all specimens. The mean cooling rate achieved from natural air (R1) to forced air (R2) is increased by a factor 2 approximatively whereas water quenching in a tank (R3) lead to cooling rate in the range of 900°C/s. Delayed (D1-D5) and interrupted (I1-I4) quench provide intermediate mean cooling rate of 100-300°C/s, which is between forced air and water quenching rate.

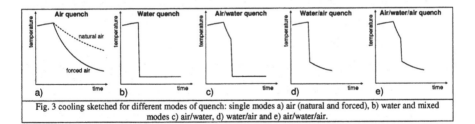

Fig. 3 cooling sketched for different modes of quench: single modes a) air (natural and forced), b) water and mixed modes c) air/water, d) water/air and e) air/water/air.

Sample N°	Condition	Qstart [°C]	Qend [°C]	Mean Cooling rate [K/s]	Bending angle after 10% [°]
Reference samples					
R1	Natural Air			-5	44 +/- 3.5
R2	Forced Air			-11	44 +/- 4.5
R3	Water quench in a tank			-970	22 +/- 3.0
Delayed quench					
D1	Water 550°C-50°C	545	43	-250	27 +/- 4.9
D2	Water 500°C-50°C	495	37	-270	29 +/- 4.2
D3	Water 450°C-50°C	445	40	-260	31 +/- 2.5
D4	Water 400°C-50°C	396	38	-290	30 +/- 3.1
D5	Water 300°C-50°C	295	42		43 +/- 4.0
Interrupted quench					
I1	Water 500°C-400°C	494	405	-130	48 +/- 3.1
I2	Water 500°C-250°C	495	251	-200	30 +/- 3.8
I3	Water 500°C-125°C	496	125	-220	27 +/- 3.7
I4	Water 500°C-50°C	495	37	-270	29 +/- 4.2
Mixed conditions					
M1	Air 500°C-400° + Water 400°C-250°C + Air 250°C-50°C	400	250		27 +/- 2.5

Table 1: Processing parameters and bending results of samples

The cooling profiles obtained by water quench are superposed in Fig. 4. It shows that for delayed quench (Fig. 4a) the ending temperature is highly reproducible while the starting temperature is controlled in less than 10°C relative to the target. For interrupted conditions (Fig. 4b), the cooling rate increases dramatically below 250°C (fig. 4b, compare I3 and I4), as a result of the Leidenfrost effect described above.

Table 1 shows that the bendability is improved by about 20° with water quenching although no significant difference in the bendability was noted between still and forced air cooling. The samples quenched with delayed quenching exhibit bending angles evolving between those 2 conditions (Fig. 5a): The bending performance remains optimum until Qstart, is equal to 400°C and then suddenly deteriorates (bending angle increases) for lower temperature. If water quench is started below 300°C, the bending behavior is similar to natural air condition (arrows).

The samples subjected to the interrupted quench experiment (Fig. 5b) do not show improvement in the bending values until that the Qend temperature approached 400°C, and then a significant improvement is observed at the Qend temperature close to 250°C. Below this critical temperature, it remains stable and similar to water quench in a tank.

The mixed cooling condition (M1) involving forced air cooling from 550 to 400°C, water quenching between 400°C-250°C and air cooling again below 250°C exhibits bendability similar to the full water quench material. This suggest that the optimum bending performance is obtained by water quenching in the critical zone of 400 – 250°C range only.

3.2 Microstructure observation

Figure 6a shows the presence of small precipitates (approx. 200 - 500 nm) on the grain boundaries of an etched sample which was air cooled below 400°C. A closer examination of the image shows the presence of two types of particles; elongated marked A and nearly spherical marked B in Fig. 6a. The size of these particles is too small for analyze via conventional spectroscopy. However, with respect to the etching reaction of the precipitate in HF reagent, it can be suggested that the nearly spherical precipitates are Mg_2Si (dissolved) whereas the un-etched elongated particle are Si [9]. Further TEM microscopy needs to be applied to confirm this statement.

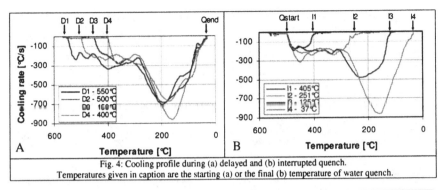

Fig. 4: Cooling profile during (a) delayed and (b) interrupted quench.
Temperatures given in caption are the starting (a) or the final (b) temperature of water quench.

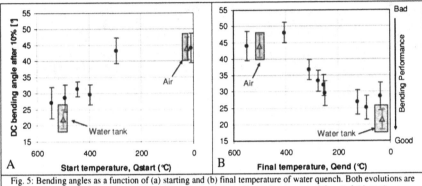

Fig. 5: Bending angles as a function of (a) starting and (b) final temperature of water quench. Both evolutions are compared with the reference conditions of natural air cooling and water quench in a tank (Arrow)

The amount and size of the particles A was influenced by the quenching conditions: For air cooled samples below 400°C (Fig. 6a), high amount of Si decorated the grain boundary whereas none of them were observed in the samples water quenched to 125°C, as seen in Fig 6d. Fewer and smaller precipitates were observed on the grain boundary in the samples which were interrupted quenched at 300°C, see Fig. 6c. The Mg_2Si precipitate marked B were always observed but in smaller quantities in water quenched samples.

From the results of this study, it is clear that the inferior bendability is related to precipitation of particles on the grain boundaries, which is consistent with the results of Himuro et al. [5]. However in this case two types of precipitates were observed and it is thought that the presence of Si particles is more detrimental to bendability than Mg_2Si. Indeed, a 6xxx alloy with a Si/Mg ratio of 2.75 has been tested here compared to a Si/Mg ratio of 1.65 in [5]. This leads to an increase of the Si-phase stability [3] and of the volume fraction of the precipitates. Thermodynamical calculations, not shown here, suggest that with the lower Si/Mg ratio, Mg_2Si is more stable than Si whereas it is the opposite for the higher Si/Mg ratio of AA6016.
Further bending analysis and microstructure examinations of 6xxx alloys with various Si/Mg ratio will be performed to clarify this.

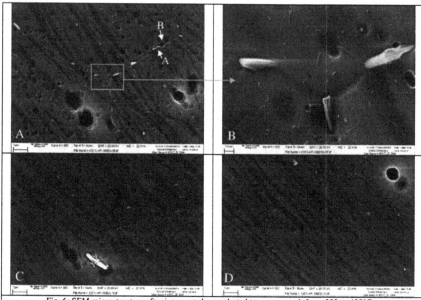

Fig. 6: SEM microstructure after interrupted quench. (a,b) water quench from 550 to 400°C
(I1); (c) water quench from 550 to 300°C (I5), (d) water quench from 550°C to 125°C (I3)

5. Conclusions

The bending performance of a 6xxx alloy with high Si excess (AA6016, Si/Mg ratio ~ 2.75.) was improved by water quenching following solutionizing heat-treatment. In this study, air and water quenching sequences were applied with various start and end temperature of water cooling. A critical temperature range for optimum bending behavior was found with an upper and a lower limit of ~400°C and ~250°C respectively. Microstructure observations suggested that bending is deteriorated by grain boundary precipitation of Si and Mg_2Si.

Acknowledgments

The authors would like to thank Novelis for permission to publish this work and to Ms L. Savioz for SEM observations.

References

[1] A.K. Gupta, D.J. Lloyd, S.A. Court, Mat. Sci. and Eng., A316 (2001) 11-17
[2] X. Wang, S. Esmaeili, D. Lloyd, Met. and Mater. Trans. A, 37A (2006) 2691-2699
[3] D.J. Chakrabarti, D. E. Laughlin, Progress in materials Science 49 (2004) 389-410
[4] M. Murayama and K. Hono, Acta mater. 47-5 (1999) 1537-1548
[5] Y. Himuro, Z. Yong, K. Matsuda, S. Ikeno, Materials Forum 28 (2004) 464-469
[6] K. Takata, K.Ushioda, M.Kikuchi, Materials Science Forum 519-521 (2006) 233-238
[7] M. Asano, S. Uchida, H. Hyoshida, Journal of Japan Institute of Light metals, 52- 10 (2002) 448-452
[8] Bending test, European Normen, DIN EN ISO 7438
[9] Aluminum and Aluminum Alloys, ASM specialty handbook, edited by J.R. Davis and Associates, ASM international, third edition, 1994

94

Aluminum Alloys: Fabrication, Characterization and Applications II
Edited by: Weimin Yin, Subodh K. Das, and Zhengdong Long
TMS (The Minerals, Metals & Materials Society), 2009

THE EFFECT OF MICROSTRUCTURE ON THE SURFACE FINISH OF EXTRUDED 6262 ALUMINUM ALLOY BILLET

Qingyou Han[1]

[1]Department of Mechanical Engineering Technology, Purdue University
401 N. Grant Street, West Lafayette, IN 47907, USA

Keywords: Aluminum Alloys, Extrusion, Solidification, EDXS, and Intermetallic Phases

Abstract

6262 aluminum alloy is essentially nominal 6061 alloy with additions of lead and bismuth for improved machinability. However, the hot extruded 6262 alloy products suffer a poor surface quality, which varies from ingot to ingot. The intent of this study focused on the effect of microstructure on the surface quality of the extruded products. The microstructure of the extruded samples was characterized and the microstructure of a sample with a good surface was compared with those of poor surface quality. To our surprise, the extruded samples contained a large number of Mg_3Bi_2 particles rather than lead-bismuth particles. The microstructure of a sample with a good surface was compared with those of poor surface quality. The volume fraction, size, and size distribution of the Mg_3Bi_2 were measured. Initial results suggest that the larger the Mg_3Bi_2 particles, the more negatively the surface quality of the extruded parts are affected.

Introduction

6262 aluminum alloy is essentially nominal 6061 alloy with additions of lead and bismuth for improved machinability [1,2]. Since lead and bismuth are soluble in the liquid alloy but insoluble in the solid primary aluminum phase, they segregate during the solidification of the alloy and form dispersed, low melting temperature small particles at the grain boundaries [3]. During machining operation, the low melting particles are melted, resulting in short and discontinuous chips [2]. Alloys that produce short and discontinuous chip during machining operation are termed as "free machining alloy" [1-5]. 2021 and 6262 are two of the major free machining aluminum alloys produced in North America [2]. Both of the alloys contain lead and bismuth.

6262 alloy is usually cast using Direct Chill (DC) casting process and hot extruded into the final products. The hot extruded 6262 alloy often suffers a poor surface finish. The yield of the extruded 6262 alloy products is only about 55%, much lower than that of the 6061 alloy products, which has a yield of over 80%[6]. The surface quality of the extruded products varies from ingot to ingot even though the composition of the alloy is within its chemical specification. The poor surface finish of the extruded alloy is due to material buildup at the die bearing surface that scratches the surfaces of the extruded billet. Little information is available in the literature on the surface finish of the extruded 6262 alloy products.

The intention of this study was to characterize the microstructure of the extruded specimens. The microstructure of a good surface quality sample has been compared with those of poor surface quality. This article reports initial experimental results on the microstructure of the extruded 6262 aluminum alloy. Possible mechanisms by which the microstructure affects the surface finish of the extruded 6262 alloy products are suggested.

Microstructure Characterization

Three specimens have been characterized. The surface quality and shape of the billets from which the specimens were taken and their chemical composition are given in Table 1. These billets were extruded from direct chill (DC) ingots of 22.5 cm (9 in.) in diameter. Before extrusion, the ingots were homogenized at 554.4°C (1030°F) for three hours and cooled at 65.6°C/hour (150°F/hour) to room temperature. The ingots were then heated to 505°C (941°F) and extruded to the final products shown in Table 1.

Table 1. The surface quality and shape of the billets and their composition (wt%).

No.	Surface Quality	Billet Shape	Billet Size	Mg	Si	Pb	Bi
1	Good	Hexagon	3.00 cm (1.25″)	0.82	0.71	0.44	0.31
2	Blemished	Hexagon	2.54 cm (1″)	0.86	0.72	0.56	0.44
3	Blemished	Round	2.54 cm (1″)	0.86	0.70	0.46	0.40

The microstructure of the as polished specimens shown in Figure 1 comprised of a dark colored particle phase and a light colored matrix of 6061 aluminum alloy. On the surface perpendicular to the extrusion direction, the particles were spherical and their sizes were non-uniform. Most of the particles were small but there were also a few particles larger than 10 μm. The particles were elongated on the surface parallel to the extrusion direction. A comparison of the microstructure of a specimen having a good surface quality with that of the blemished specimens is shown in Figure 1. The sample with a good surface finish, shown in Figure 1(a), appears to have fewer and smaller large particles than the samples with a poor surface finish shown in Figure 1(b) and 1(c).

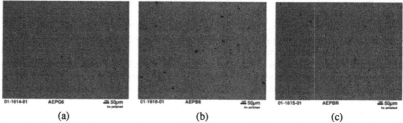

| (a) | (b) | (c) |

Figure 1: The microstructure on the as polished surfaces of (a) sample number 1(good surface finish), (b) sample number 2(poor surface finish), and (c) sample number 3(poor surface finish). The photographs were taken on the surfaces perpendicular to the extrusion direction of the billets.

Image analysis was then carried out to determine quantitatively the difference of the particles between the sample having a good surface finish sample and those having a poor surface finish. Over seven thousand particles in each sample were measured using an image analyzer on the surface perpendicular to the extrusion direction of the billets. The particle size distributions in those three samples, shown in Figure 2, were very similar for the small particles. Over 95% (number percent) of the particles were smaller than 5 μm. For particle size smaller than 5 μm,

96

the three curves are almost overlapped. Difference exists for large particles but the difference is not evident in the distribution curves since the population of the large particles is much smaller than those of small particles.

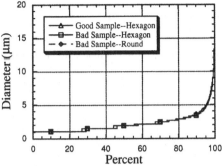

Figure 2: The particle size distributions in the hot extruded 6262 aluminum alloy samples.

The volume fraction and mean size of the particles were also examined. The sample with a good surface finish contained a smaller volume fraction of particles with a slightly smaller mean particle size than that in samples with a poor surface finish. The difference between the sample having a good surface finish and those having a poor surface finish was mainly due to the difference of large particles in the specimens. Thus it makes more sense to compare the large particles in both types of specimens.

The maximum particle size in each field under optical microscope was measured. The maximum particle size is defined as the largest particle size in each field of 500×700µm. The fields were randomly chosen over the cross-section of the specimen perpendicular to the extrusion direction of the billet. The results are shown in Figure 3. The maximum particle size in a sample having a good surface finish was in the range between 10 and 20µm, whereas that in a sample having poor surface finish was in the range between 10 to 33µm. The particle size in the samples with a poor surface finish could be as large as two times of that in the sample with a good surface finish.

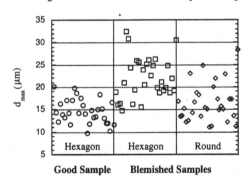

Figure 3: The maximum particle sizes in the hot extruded 6262 aluminum alloy samples. The maximum particle size is defined as the largest particle size in each field of 500×700µm.

Discussion

The results of microstructural characterization indicate that the size and the number of the large particles are the major differences between the specimen with a good surface finish and those with a poor surface finish. The specimen with a good surface finish appears to have smaller and fewer large particles than specimens with a poor surface finish. This can be related to the chemical compositions of the specimens. As it is given in Table 1, the concentrations of lead and bismuth in the specimen with a good surface finish are lower than those in the specimens with a poor surface finish. Higher lead and bismuth concentrations lead to higher volume fractions of the particle and possibly larger particle size in the extruded products. However it is difficult to believe that the large particles of lead and bismuth can damage the surfaces of the extruded products since lead and bismuth are soft materials. Besides, the volume fraction of the particles seems much larger than that can be formed in a specimen with slightly higher lead and bismuth contents in the alloy. In order to find the answers to this issue, the particles were further analyzed using energy dispersive X-ray microscopy (EDXS).

EDXS analysis was carried out on a numerous particles. Figure 4 shows a typical result. The EDXS analysis revealed that the particles in the extruded specimens were mainly composed of bismuth and magnesium, rather than lead and bismuth that are supposed to form low melting point Pb-Bi eutectic particles for improved machinability of the alloy. The magnesium and bismuth concentrations in the particles were about 51.56 and 43.79 atomic percent respectively. Lead concentration in the particles accounted for only 0.49 atomic percent. This was quite a surprise since lead and bismuth were intentionally added in the alloy in order to form lead-bismuth eutectics that melt at 125.5°C [7]. The lead content in the binary lead-bismuth eutectic should be around 55 atomic percent [7]. Thus the particles in the extruded specimens were not lead-bismuth eutectics at all.

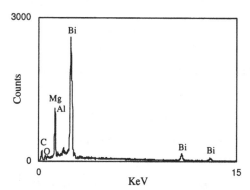

Figure 4: An EDXS analysis of the particles. It reveals that the particles in the extruded specimens are mainly composed of bismuth and magnesium, rather than lead and bismuth that are supposed to form low melting point lead-bismuth eutectic particles for improved machinability of the alloy.

98

The phase diagram of magnesium and bismuth is illustrated in Figure 5 [7]. Mg_3Bi_2 is the only phase that exists in the alloy system apart from the magnesium and bismuth phases. The content of bismuth in Mg_3Bi_2 is around 40 atomic percent, which is very close to the concentration of bismuth in the particles of the extruded specimens. This suggests that the particles shown in Figure 1 are Mg_3Bi_2 particles rather than the lead-bismuth eutectic particles. Figure 5 also shows that the melting point of the Mg_3Bi_2 phase is 821°C (1510°F), much higher than the melting point of the aluminum alloys. Thus, the formation of Mg_3Bi_2 is most likely related to the poor extruded surface quality, since the Mg_3Bi_2 is a hard intermetallic phase with a high melting temperature. If these particles are trapped at the die bearing surface during extrusion of the 6262 alloy, they are able to blemish the surfaces of the extruded products.

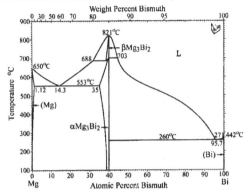

Figure 5: The phase diagram of magnesium-bismuth binary alloys.

Assuming that the particles are mainly comprised of Mg_3Bi_2, we are able to explain the experimental results. Surfaces of the extrusion product containing a higher volume fraction and larger size of particles are more easily to be blemished. As illustrated in Figure 3, the maximum particle size in samples with a poor surface finish is two times as high as that in the sample with a good surface finish; the volume fraction of particles in samples with a poor surface finish is 20% higher than that in the sample with a good surface finish. The large size and high volume fraction of the large particles account for the poor surface quality of the samples. It appears that the larger particle size and higher particle volume fraction are related to the higher concentrations of magnesium, bismuth and lead in the alloy. Therefore, in order to achieve a good surface quality of the extruded products, the chemical composition of the alloy has to be controlled.

Apart from the detrimental effect of the Mg_3Bi_2 on the surface quality of the extruded products, the formation of Mg_3Bi_2 phase consumes magnesium which is added for age hardening and bismuth that is added for the improved machinability. Thus it could be detrimental to the mechanical properties as well as the machinability of the alloy. More research is needed to investigate the conditions under which the Mg_3Bi_2 particles form and grow during the processing of aluminum 6262 ingots.

Conclusions

The microstructure of a specimen with good surface quality is compared with those of poor surface quality. The following conclusions have been made:

(1). 6262 aluminum alloy extruded at 505°C (947°F) contains a large number of Mg-Bi particles, most likely Mg_3Bi_2 particles. If these hard intermetallic Mg_3Bi_2 particles with a high melting point are caught at the die bearing surface, they are likely to blemish the surface of the extruded products.

(2). The sample with a good surface finish has a lower volume fraction and smaller size of particles than samples with a poor surface finish. This is partly due to the fact that the magnesium, lead, and bismuth content in the sample with a good surface finish is lower than that in the samples with a poor surface finish.

Acknowledgment

This research was supported by the U.S. Department of Energy, Assistant Secretary for Energy Efficiency and Renewable Energy, Office of FreedomCar and Vehicle Technologies, Automotive Lightweighting Materials Transportation Technology Program, the ORNL SHARE User Facility, Office of Industrial Technologies, Industrial Materials for the Future (IMF) Program, Materials Processing Laboratory Users (MPLUS) Facilities, under contract DE-AC05-00OR22725 with UT-Battelle, LLC. The authors would like to thank J.R. Mayotte for optical metallography, L.R. Walker for the EDXS, and M.L. Santella and T.J. Huxford for reviewing the article.

References

1. S. Sircar, "X6030, A New Lead-Free Machining Alloy," *Materials Science Forum*, vols 217-222 (1996), pp. 1795-1800.
2. M.C. Roth, G.C. Weatherly, and W.A. Miller, "The Role of Liquid Metal Embrittlement in Free-Machining Alloys Containing Low-Melting-Point Inclusions," *Canadian Metallurgical Quarterly*, vol. 18 (1979), pp. 341-248.
3. D.W. Davis, "Machinability and Microstructure of Some Common Non-Ferrous Metals and Alloys," *Metals Technology*, May-June, 1976, pp. 272-287.
4. M.E. Hydon, "Machinability," London, The Iron and Steel Institute, 1967, p.185.
5. C.M. Spillard, "UltraAlloy 6020: A Lead-Free Aluminum Alloy Featuring 'A' Rated Machinability," SAE Special Pubilcations, vol. 1350, Aluminum in Automotive Applications, 1998, pp. 61-68.
6. Q. Han and M. Shelley, "Improving the Surface Finish of Extruded 6262 Alloy Billet," MPLUS Report, Oak Ridge National Laboratory, Oak Ridge, TN, USA. September, 2002.
7. H. Baker et al, (Eds.), ASM Handbook, vol.3, Alloy Phase Diagrams (Materials Park, OH: ASM International), 1992.

DEFORMATION TEXTURES AND PLASTIC ANISOTROPY OF AA6XXX AT WARM TEMPERATURE

M. Ghosh[1,2], A. Miroux[1], J. Sidor[1], L. Kestens[2]

1Materials innovation institute; Mekelweg 2; Delft, 2628 CD, the Netherlands
2Delft University of Technology/3ME/MSE; Mekelweg 2; Delft, 2628 CD, the Netherlands

Keywords: warm forming, Al-Mg-Si alloy, earing, r-value, texture

Abstract

Tensile and plane strain compression tests as well as deep drawing tests have been used to investigate the forming behaviour of Al-Mg-Si alloys from room temperature to 250°C. In addition to the expected reduction of yield strength with increasing temperature, it is found that temperature also significantly influences the plastic anisotropy of the sheets. The earing profile of drawn cups shows a four-fold symmetry after drawing at room temperature, while at higher temperature the earing profile presents a 2-fold symmetry. The analysis of the deformed microstructures shows that other slip systems than {111}<110> can be activated at higher temperature. Crystal plasticity calculations reveal that for an adequate combination of {hkl}<110> slip systems a good correspondence between the experimental and calculated r-value and textures at different temperatures is obtained.

Introduction

Compared to low carbon steel sheets the application of aluminium sheet alloys are heavily restricted in automobile and automotive industry owing to their higher cost and lower room temperature formability in spite of their light weight and high corrosion resistance. The tensile elongation of aluminium alloys is considerably lower than the value for aluminium-killed steel [1] (30% compared to 50% for A-K steel). In terms of forming behavior aluminium alloys exhibit a forming limit strain of about 25% compared to about 45% for A-K steels under a plane strain condition at room temperature [2]. Various investigations reported that the room temperature formability can be improved by increasing the forming temperature [3,4]. Formability of sheet metal depends on both intrinsic (microstructure, precipitate, constituent particles etc.) and extrinsic (temperature, strain rate etc.) parameters. In course of optimizing the alloy screening tests are very useful. These screening tests include uniaxial tensile, plane strain compression and deep-drawing. Uniaxial tensile test is the most common and popular among the all screening tests principally because of its easier control over the test parameters. Although it is not possible to interpret the material behavior during biaxial or more complex loading conditions in terms of tensile test results, it is important in terms of estimating various parameters that influences formability [5]. Plane strain compression tests are performed in order to achieve the large strain deformation response of the materials at elevated temperatures. They are useful in developing the constitutive equations used for finite-element (FE) simulations of hot forming operations. Deep-drawing tests are done in order to add the complexity in estimating the formability at higher temperature and also to introduce the problem of plasticity with temperature.

As part of an investigation on improvement of formability this paper reports results on the mechanical behavior of the aluminum alloy EN AW-6016 from room temperature to 250°C by tensile, plane strain compression (PSC) and deep-drawing tests. The true plastic stress-strain curves generated from tensile and PSC or load-displacement plot from deep-drawing have been used to characterize the work-hardening behaviour at room and 250°C. The earing profile obtained from the deep-drawing cups and r-values from the tensile tests were used as a measure of anisotropy. Electron Backscattered Diffraction (EBSD) maps has been measured to identify the textural changes and tracing of crystallographic planes activated during deformation both at room temperature and 250°C. The slip planes so identified have been used to calculate the final texture using a Visco-Plastic Self Consistency Model (VPSC) [6].

Experimental Procedure

The 6xxx aluminium alloy AA6016 (EN AW-6016) has been used for the present investigation. The material was cold rolled, solutionised and naturally aged (T4). The final sheet thickness is 1.0 mm. The alloy contains 0.42 Mg, 1.03 Si, 0.06 Cu and 0.25 Fe in wt%.
The tensile, PSC and deep-drawing experimental procedures along with description of the texture measurements by XRD and EBSD have been described elsewhere [7]. The orientation of the deformation bands visible on the EBSD maps of the deformed samples were compared to the local crystallographic orientation to determine which crystallographic planes are containing the bands.

Results

Figure 1.a represents the plastic true stress-true strain curve obtained from tensile tests at two temperatures (22, and 250°C) and at $0.1s^{-1}$ strain rate. At room temperature materials have higher point of yielding with larger rate of work-hardening compared to higher temperature (250°C). The higher yield point at room temperature compared to higher temperatures is the result of two effects: the material softens during the 30 s holding at 250°C preceding drawing as static ageing curve shows a decrease of hardness from 81Hv to 63Hv and also due to the effect of temperature on yield point. Figure 1.b shows the drawing force evolution. The strength of the materials and rate of work-hardening becomes lower with increasing drawing temperature from room temperature to 250°C. It is also depicted that there has been almost no difference in flow curves for tensile tests along RD, TD and 45° directions at room temperature. At 250°C, 6016 specimen showed the clear effect of anisotropy, with tensile test along 45° having higher work-hardening rate than along RD and TD.

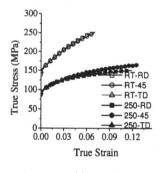
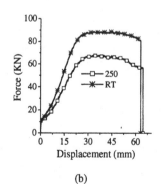

(a)   (b)

Figure 1. Strain-hardening response at room temperature and 250°C during (a) tensile tests along different directions and (b) deep drawing.

102

Figure 2.a and b show that the deformation temperature strongly influences the normal anisotropy and earing behaviour. At room temperature, r-values show a minimum at 30° to the rolling direction and the earing profile exhibits four-fold symmetry. When the deformation temperature is increased to 250°C, r-value increases for all directions but the earing profile nearly exhibits a two-fold symmetry.

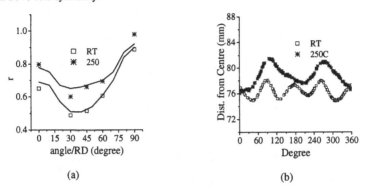

(a)   (b)

Figure 2. (a) r-value measured by tensile tests (symbols) and calculated by the VPSC model (line) and (b) drawn cup foot-print representing the earing profile at room temperature and 250°C.

The orientation distribution function (ODF) of the AA6016-T4 sheet shows a texture dominated by Cube and Goss grains and a η-fibre ($\varphi_1 = \varphi_2 = 0°$) (Figure 3). Figure 4.a and b show the texture in the flange along the rolling direction after deep drawing at room temperature and at 250°C. It can be well noticed that the texture changes after deformation while the textures obtained after drawing at both temperatures are qualitatively similar. As-received texture with similar Cube and Goss volume fraction shows reduction in Cube and increment in Goss and P after deformation at both the temperatures. Both textures after deformation have the maximum close to Goss and P with an orientation spread along the α-fibre ($\phi = 45°$, $\varphi_2 = 0°$) and towards Cube. There is also a η-fibre between Goss and Cube.

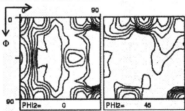

Figure 3. ODF of 6016-T4 sheet calculated from XRD measurement.

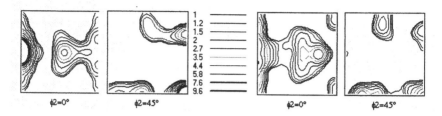

Figure 4. ODF in the flange after drawing at room temperature (a) and 250°C (b) calculated from EBSD measurement

Figure 5.a and b represent the ODF after PSC at room temperature and 250°C. Like deep-drawing these textures are also qualitatively similar. The α-fibre from Goss to I-Goss is also found to be discontinuous. Compared to the as-received material the intensity of the cube component decreases in a similar way than during deep drawing and the increase of S and R is even more pronouced during PSC compared to deep-drawing. The textures do change after deep-drawing and plane strain compression but no important differences can be observed between textures after room temperature and warm temperature deformation.

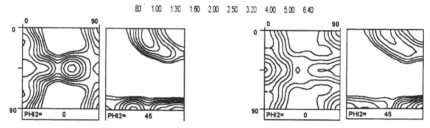

Figure 5. ODF of the specimens after PSC at a) RT and b) at 250°C

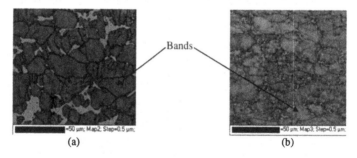

Figure 6. Band contrast map (a) room temperature and (b) 250°C used to identify the crystallographic slip planes analyzing the direction of the deformation bands

From the band contrast map obtained by EBSD deformation bands have been identified and the possible crystallographic planes which may contain these bands have been calculated. It has been observed that at room temperature {111} slip plane is the only solution while at 250°C combination of {111}, {110}, {100} and {111} slip planes are possible solutions.

Discussion

The presence of the four-fold symmetry after room temperature deep-drawing and the planar anisotropy measured at room temperature are typical of materials with Cube texture [8]. With increasing the deformation temperature to 250°C a remarkable change of the plastic anisotropy in the material has been observed. The four-fold symmetry of the earing profile transforms to a two-fold symmetry while the r-values increase uniformly. On the contrary, temperature was found to have limited effect on the deformation textures. Although the compression direction is along TD during deep drawing and along ND during PSC, the intensity of the cube orientation decreases significantly at both temperatures and for the two deformation modes. As the textures have been calculated from EBSD measurements with relatively poor statistical accuracy (typically 120 grains) and the compression directions are different for the two mechanical tests the other differences observed between the deformed textures at room temperature and at 250°C after deep drawing and PSC are more difficult to interpret.

The experimental r-values have been simulated by the VPSC model using {111} as the only slip plane at room temperature and four families of slip systems {111},{110},{100} and {112}<110> at 250°C with the CRSS ratios 0.9:1:1.1:1 [9] (figure 2a). The results at room temperature compare very well with the experimental measurements while at 250°C model predicts slightly higher values at 30° and 90°. Using the right CRSS ratios for the four families of slip systems it is possible to calculate r-values at 250°C quite efficiently. These slip systems have been used as input for the VPSC model to simulate the texture after 35% deformation during PSC after discretisation of the initial texture in 2000 orientations (figure 6).

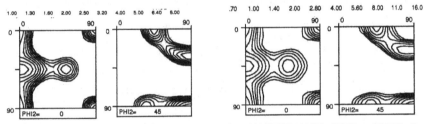

Figure 6. ODF from the VPSC model using a) {111} and b) {111}, {110}, {100} and {111} slip systems in PSC mode with 35% deformation

The textures generated with 1 or 4 slip systems are qualitatively very similar. They are also very much similar to the experimental ones. The model predicts well the decrease of the cube component intensity although it underestimates this decrease. The other main quantitative differences between the modelled and experimental deformed textures are the higher intensity of Cu and the lower intensity of I-Goss in the modelled textures and the position of maximum along the α-fibre, which is around the Br component in the modelled texture and closer to P orientation in experimental results. Nevertheless it can be concluded from the above results that although deformation textures are similar at different temperatures the activation of different families of slip systems is still possible, which may directly affect the r-value profile.

Conclusion

A remarkable change in plastic anisotropy during deep-drawing is observed when the temperature is increased from room to 250°C. The earing symmetry transforms from four to two fold and the r-value increases while keeping the same profile. However textural changes do not reflect the changes in earing behaviour. Texture changes during deep-drawing and PSC compared to the initial texture but they are hardly affected by temperature. The r-value and texture predicted by the VPSC model using different combinations of slip systems for room temperature and 250°C deformation show good match with the experimental results.

Acknowledgement

This research was carried out under the project number MC4.02106 in the framework of the Research Program of the Materials innovation institute M2i (www.m2i.nl), the former Netherlands Institute for Metals Research. All the partners of this project are highly acknowledged for their collaboration and support.

References

1.   Aluminium Standards and data 1993, The Aluminium Association, Washington, D. C., (1993).

2.   Beaver P.W. and Parker B.A., (Proceedings of the 12th Biennial Congress of the International Deep-Drawing Research Group, S. Margherita Ligure, Italy, 1982), 85.

3.   Morris L.R. and George R.A., "Warm forming high-strength Aluminium parts", (International automotive engineering Congress and Exposition, *Society of Automotive Engineers*, Cobo Hall, Detroit, MI, 1977), paper 770206, 1-9.

4.   O'Donnell R.G. and Parker B.A., "Parameters affecting the strain rate sensitivity of aluminium alloys", (Proceedings of International deep-drawing research group 13th Biennial Congress, Melbourne, Vic., 1984), 372-382.

5.   Ghosh A.K., "Strain localization in diffuse neck in sheet metals", *Metallurgical Transaction*, 5A, (1974), 1607-1616.

6.   Lebensohn R.A. and Tome C.N., Acta *Metallurgica and Materialia*, 41 (9) (1993), 2611-2624.

7.   Ghosh M., Miroux A., Werkhoven R., Bolt P.-J., Sidor J., and Kestens L., Textural Changes after Warm Temperature Forming of AA6xxx, (Proceedings of the International Conference of Aluminium Alloys (ICAA-11), 2008), 782-787.

8.   Engler O. and Hirsch J., *Materials Science and Engineering A*, 452-453, (2007), 640-651.

9.   Bacroix B. and Jonas J.J., Textures and Microstructures, 8-9, (1988), 267-311.

Aluminum Alloys: Fabrication, Characterization and Applications II
Edited by: Weimin Yin, Subodh K. Das, and Zhengdong Long
TMS (The Minerals, Metals & Materials Society), 2009

Formation of the {111}<110> and {111}<112> shear bands and the <111> fiber texture during moderate and heavy wire drawing of 5056 Al-Mg alloy

Mohammad Shamsuzzoha[1,2], Fingling Liu [1]

[1] Department of Metallurgical and Materials Engineering.
[2] School of Mines and Energy Development, The University of Alabama, Tuscaloosa, Alabama

Keywords: Wire drawing, Fiber texture, Deformation bands

Abstract

Metallography, conventional transmission electron microscopy and x-ray diffraction techniques have been applied to study the defect microstructure of moderately and heavily wire drawn 5056 Al-Mg alloy. Samples drawn moderately (~70% of the original value) has been found to be comprised of columnar grains with no evidence of any recrytallization and contain deformation bands. Deformation bands are made of closely spaced parallel slip bands, which lie on {111} and extends along a <110> and contribute to the development of a moderate <111> fiber texture. These samples also show material decohesion normal to fiber axis. The matrix of the sample drawn heavily (~ 40% of the original value) showed fibrous microstructure with very little evidence of recrystallization, but exhibit a strong <111> texture. Deformation bands in these samples have been found to lie on {111} planes but extend along <112>.

Introduction

In polycrystalline structure of metals and its alloys constituent grains usually assume random orientations. During metal forming, such as wire drawing, these grains undergo plastic deformation, in which constituent crystals rotate towards a stable orientation. The deformation process thus leads to the development of some non-randomness of the grain orientation, known as preferred texture. The nature of texture depends upon the type of material, for example drawn wires of the f.c.c metals generally exhibit a dual <111> and <100> texture [1-2]. Materials properties of single crystals vary with the crystal planes and direction, therefore, development of texture in the polycrystalline aggregate usually exhibits anisotropic materials properties. Such texture and associated materials anisotropy may or may not be suitable from the view point of materials' application In order to fully utilize the property of anisotropy to advantage, it seems very desirable to custom make a texture for a specific application. Such a manufacturing of texturing in materials can only be easily achieved, if structural mechanisms that undergo during the metal forming process of wire drawing are known. Understanding of structural mechanisms requires extensive study of the microstructure and crystal structure of samples wire drawn to various level of strain. The present work was undertaken to unravel the necessary structural information in relation to the fiber texturing in a 5056 aluminum alloy. This paper describes the various crystal and microstructural aspects of the formation fiber texture in the alloy during wire drawing.

Experimental Procedures

Material used in the present study is a 5056 aluminium alloy, which was purchased from ALCAN (Aluminium Company of Canada), and supplied in the form of cylindrical rods of 9.525 mm diameter. Exact composition of this alloy is given in Table1. The alloy was reduced in two batches by a single pass wire drawing to 6.858 mm and 3.962 mm to give close to 30 and 60 percent reduction respectively.

The microstructure of as received alloy rods as well as of drawn wires was examined by optical microscopy, transmission electron microscopy and by x-ray diffraction techniques. TEM micrographs and electron diffraction patterns were taken of the drawn samples using a Hitachi H-8000 transmission electron microscope operating at 200 keV. For preparation of usual and longitudinal section thin foil specimens for electron microscopy, cylindrical discs of 3 mm were cut using an ultrasonic disc cutter. The discs thus prepared were sliced with a diamond saw to o.5 mm thickness or less, then electropolished in a solution of 70% methanol and 30% nitric acid at a temperature of -23$^\bullet$ C and at a voltage of 14 V. Polishing was halted just after perforation. The specimens were then cleaned in distilled water, ethanol, and ecetone. Pinhole x-ray diffraction transmission technique was used for the determination of texture in the wire drawn sample. For determination of fiber texture, the wire was placed in a transmission pinhole camera with its axis parallel to one edge of the rectangular flat film. –ray exposure was taken on a standard front-reflection camera with specimen –film distance kept at 3 cm. Ni-filtered Cu radiation generated at 40 kV/20mA was employed through a very fine collimator of diameter of 0.075mm for scattering. The exposure time for each photograph was 5 hours. For determination of fiber texture, the x-ray data   measurement procedures described by Klug and Alexander [3] was adopted

Result and discussion

Figure 1a and 1b shows the longitudinal and cross-sectional view optical micrographs of as received alloy. The micrographs exhibit that the bulk matrix of as received alloy is composed of equiaxed grains with an average size of 28µm. Transmission electron microscopy of the alloy also revealed the equixed microstructure as evident in the bright field micrograph (Figure 2a) of the alloy matrix. The matrix in this micrograph also exhibit precipitates of various shapes. Electron diffraction patterns taken of these precipitates revealed that a vast majority of the precipitates belonging to the Al_3Mg_2 phase. The pinhole x-ray diffraction pattern taken of the rod axis of the alloy is shown in figure 2b. The pattern shows the existence of {111} and {200} Debye rings, indicating that the alloy has randomly oriented grains in its microstructure.

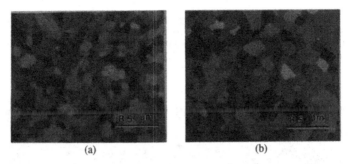

(a)   (b)

Figure 1 Optical micrographs of the matrix in the as received alloy, (a) longitudinal section, (b) cross-sectional section.

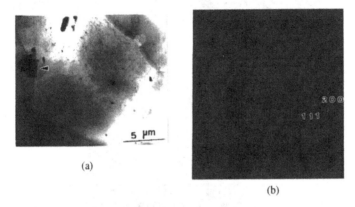

(a)

(b)

Figure2. (a) Bright field TEM micrograph showing the grain structure in the matrix of as received alloy. Arrow indicates the location of a typical precipitate in the micrograph.
(b) Pinhole x-ray diffraction pattern taken of the alloy.

The microstructure of the moderately drawn (~ 30% reduction, at which the alloy did not exceed its flow stress) wire revealed markedly different grain structure than that present in the as received alloy. Figure 3(a) and 3 (b) show the longitudinal and cross section optical micrographs of the 30-% drawn wire samples respectively. The longitudinal section micrograph exhibit longer grains oriented more along the wire axis with very little difference in the size between the edge and the center. Cross sectional section micrograph shows no special for the constituent grains and can generally be described as of the microstructure containing grains of irregular morphology. Pinhole x-ray diffraction pattern (figure 3b) of the moderately drawn sample always exhibit two very faint and incomplete Debye rings, each of which includes four intense equally spaced Debye arcs (elongated diffraction spots). The presence of incomplete Debye rings in the pattern indicates that the grains in the alloy are not randomly arranged. The existence of the elongated diffraction spots in the Debye rings indicates that certain crystal planes in the constituting grains of the investigated sample take lead in the contribution of the diffraction pattern. The angle between the Debye arc and the axis of the photographic plate as shown in the photo and the corresponding Bragg's angle of the Debye ring, upon measurement described by Clog and Alexander[3] revealed that the wire has a moderate <111> fiber texture.

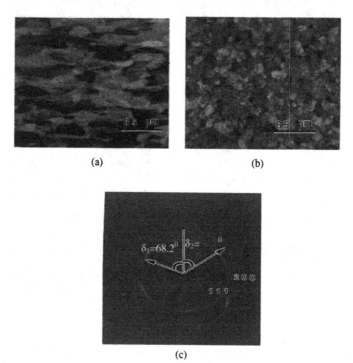

(a)   (b)

(c)

Figure 3 (a) and (b) Longitudinal and cross sectional Optical micrographs of the matrix in the moderately drawn wire. (c) Pinhole x-ray diffraction pattern taken of the moderately drawn sample

Under the TEM wire drawn samples also exhibited columnar grains and no recrystallized microstructure. Figure 4a shows a low magnification TEM micrograph of a columnar segment present in the longitudinal section of the wire. The micrograph exhibits the presence of shear bands, some of which is macroscopic in that they extend across a number of grains of the as received sample.

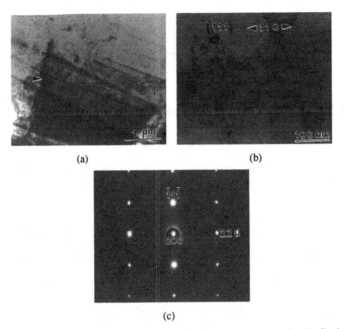

(a)   (b)

(c)

Figure 4. Low magnification TEM micrographs (a, b) of bulk matrix along longitudinal view of the moderately drawn sample. Interface between two neighboring grains is marked by an arrow.(c) Electron diffraction pattern taken with electron beam parallel to [-112]of neighboring deformation bands in (a).

The axis of these slip bands was found to lie between 30° to 40° with the drawing axis, High magnification TEM micrograph (figure4b) from such a group of shear bands also revealed that individual shear band is comprised of a large number of micro and nano bands that are also closely parallel to the average axis of the constituent shear band.

micro and nano bands revealed that these micro and nano bands lie closely on a {111} planes and extend close to <110> directions. Slip bands has been found [4] to have a {111}/<110> growth habit in the f.c.c system, hence occurrences of a similar growth habit for the micro and nano bands in the drawn wire allows them to be the slip bands. This implies that an individual shear band in the wire is formed by aggregation of slip bands. The pinhole x-ray diffraction pattern (figure 5) of the moderately wire drawn sample always revealed very faint and incomplete Debye rings, indicating that the grains in the alloy is not randomly arranged.

Figure 5. TEM micrograph showing a crack passing through many entrapped precipitates in the moderately drawn sample.

Precipitates of Al_3Mg_2 phase, present in as received samples are highly brittle [5] and, therefore, do not stain as easily as matrix. Wire drawing of these samples is expected to leave more residual stress on precipitates than the surrounding matrix. The relatively high strength of endurance to straining by precipitates makes them ideal site from which material decohesion can occur. In the present study cracks found lying at the interface between deformation bands of the drawn wires, often exhibited (Figure 5) entrapped precipitates. It is believed that the entrapped particles present in the path of crack played a major role in the development of cracks either by nucleating the initial materials decohesion or by facilitating the growth of crack.

The microstructure of the heavily drawn (~ 60% reduction, at which alloyexceeds its flow stress) wire revealed more defective grain structure than that present in the moderately drawn wire. Longitudinal micrograph of the wire exhibited fibrous microstructure (Figure 6a) typical of heavy metal forming. In cross-sectional view (Figure 6b) the microstructure appears to comprise of an ensemble of extremely small grains of irregular morphology. Pinhole x-ray diffraction pattern (figure 6b) of this heavily drawn wire also exhibited two very faint and incomplete Debye rings, each of which includes four intense equally spaced, but more intense Debye arcs (elongated diffraction spots) than those found in the moderately drawn sample. The pinhole x-ray diffraction pattern, upon an analysis that is identical to that employed for the determination of texture in the moderately drawn wire yielded a strong <111> fiber texture for the drawn wire.

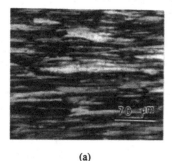

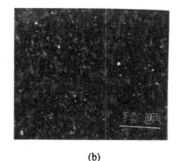

(a)   (b)

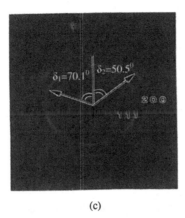

(c)

Figure 6. (a) and (b) Longitudinal and cross sectional Optical micrographs of the matrix in the heavily drawn wire. (c) Pinhole x-ray diffraction pattern taken of the heavily drawn sample.

Thin foil specimens prepared from the wire upon TEM investigation also yielded fibrous grain microstructure, and revealed little evidences of recrystallization. In longitodinal section TEM specimen of the wire, individual fiber bundle, in its proper orientaion, always exhibited constituent deformation bands. Figure 7a shows a bright field TEM micrograph of such a group of deformation bands. The electron diffraction taken of the band is given in Figure 7b. Analysis of the patterns revealed that the individual band lies on a {111} plane and extends along <112> direction. This crystallograhic characteristic of deformation band belonging to the heavily drawn sample is signifantly different from that of {111}<110> slip band fopund in the moderately drawn sample. It suggests that beyond alloys's flow stress deformation in wire drawing is accomplished by the slippage of {111} planes but along <112> directions.

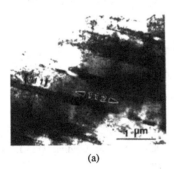

(a)

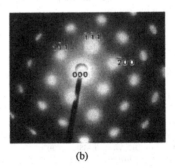

(b)

Figure 7. TEM micrographs (a) showing deformation bands in the longitudinal section microstructure of the heavily drawn sample. (b) Electron diffraction pattern taken with electron beam parallel to [011] of neighboring deformation bands in (a).

Figure 8. TEM micrograph showing deformation bands rotate around a constituent precipitate present in the heavily drawn sample.

In the course of TEM investigation of this heavily drawn sample, it was found that deformation bands frequently rotate around constituent particles (Figure8). However, interface between deformation bands were seldom found to be the site of material decohesion.

Conclusion

Determination of structural status of the bulk matrix of 5056 alloy prior to and following moderate and heavy drawings by the present study reveals an important conclusion about the deformation behavior. It reveals that the alloy attains <111> fiber texture in both moderate and heavy drawing, but the microstructure of the moderately drawn samples remaining below its flow stress exhibits {111}<110> deformation bands. Whereas microstructure of the heavily drawn samples exceeds its flow stress and exhibits {111}<112> deformation bands. These observations suggest that the formation of the <111> texture below and above flow stress occurs by different mechanisms.

References

[1] C. S. Barrett, and T. B. Massalski, "The structure of Metals" John Wiley, New York, and London (1966).
[2] S. C. Dexter, and I. G. Greenfield, Metallurgical Transaction 2, 181(1971).
[3] H. C. Klug, and L. E. Alexander,"X-ray Diffraction Procedures for Polycrystalline and Amorphous Materials", John Wiley and Sons Inc., New York, p 561(1967).
[4] G. E. Dieter, "Mechanical Metallurgy"{' McGraw-Hill, Book Company, New York, p237(1986).
[5] E. H. Chia and O. J. Tassi, "Nonferous Wire Handbook, Volume 2, Editor: Otto J. Tassi, The Wire Association International Inc. Gurlford, Connecticut,(1981).

Aluminum Alloys: Fabrication, Characterization and Applications II
Edited by: Weimin Yin, Subodh K. Das, and Zhengdong Long
TMS (The Minerals, Metals & Materials Society), 2009

POLE FIGURE CHARACTERISTICS OF ANNEALED ALUMINUM ALLOY 6061 IN DIFFERENT MAGNETIC FIELDS UP TO 30 TESLA

Samuel T. Adedokun[1], Olalere Ojo[2], Akin Fashanu[3], Bayo Ogunmola[3]

[1]ΓAMU-ΓЗU College of Engineering, Tallahassee, FL, USA
[2]Ladoke Akintola University of Technology, Ogbomoso, Oyo State, Nigeria
[3]University of Lagos, Akoka-Yaba, Lagos, Nigeria

Abstract

This work presents the changes in the pole figure characteristics of an aluminum alloy 6061 given 85% deformation by cold rolling and later heat treated at 400°C in different magnetic fields of up to 30 Tesla for different periods of time. Pieces of samples from the rolled specimen were heat treated in a resistive magnet of 30 Tesla strength with 50 mm bore. The texture changes in the samples were quantified by carrying out texture measurements through an x-ray diffractometer equipped with a texture goniometer. Changes in the texture with the use of the inverse and complete pole figures indicate that the strength of the magnetic field had no effect on the strength of the texture of the material.

Introduction

In this work, inverse and complete pole figures obtained after aluminum alloy 6061 with nominal composition of Si 0.60, Fe 0.7, Cu 0.28, Mn 0.15, Mg 1.0, Cr 0.20, Zn 0.25, Ti 0.15 and Al remainder was deformed from 13 mm thickness to 2 mm thickness through a multi-step rolling method at room temperature are presented. Aluminum alloy 6061 is widely used for construction of aircraft structures, such as wings and fuselages, more commonly in homebuilt aircraft than commercial or military aircraft. The objective of this study is to present the effects of magnetic field of strengths of up to 30 Tesla on the inverse and complete pole figures of aluminum alloy 6061 when heat treated at 400°C for different periods of time.

Experimental

A plate of aluminum alloy 6061 obtained from Copper and Brass, Inc located in Atlanta, Georgia was solution heat treated for one hour at 500°C to homogenize its grains. Thereafter, it was quenched in water at room temperature and rolled from 13 mm thickness to 2 mm using a multi-step rolling process at room temperature. Samples of average size of 2 cm x 1 cm were cut from the rolled specimen and heat treated t 400°C for 10, 20 and 30 minutes. The heat treatment (annealing) was done in magnetic field of strengths 10, 20 and 30 Tesla. Also, the annealing process was repeated for the same periods of time and at temperature of 400°C when the magnetic field was turned off. The heat treatment under the magnetic field was done with the rolling direction of each of the samples parallel to the direction of the field because this position produces maximum effect of the field on the material characteristics.

The annealed samples were cold mounted in a resin for about 8 hours, then polished down to 3 micron surface finish. Texture measurements were carried out on the polished samples using a Phillips x-ray diffractometer equipped with texture goniometer. Texture data obtained were analyzed with PC Texture© and popLA© softwares to obtain complete and inverse pole figures of the samples. Results obtained are presented below.

Results and Discussion

Pictures in Figure 1 show the complete pole figures and inverse pole figures for the rolled sample before heat treatment. The complete and inverse pole figures for the heat treated samples are also presented in Figures 2, 3, 4 and 5.

The complete and inverse pole figures in Figure 1 indicate a well textured rolled face-centered cubic aluminum alloy material with highest intensities for (111) pole. However, the pictures in Figures 2, 3, 4 and 5 show that at the onset and during the heat treatment the texture started disappearing. The magnetic field was not seen to play a significant role on the pole figures.

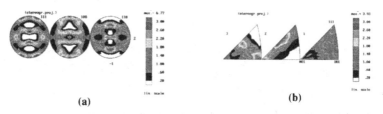

(a)   (b)

Figure 1: (a) Complete and (b) inverse pole figures for the rolled sample before heat treatment.

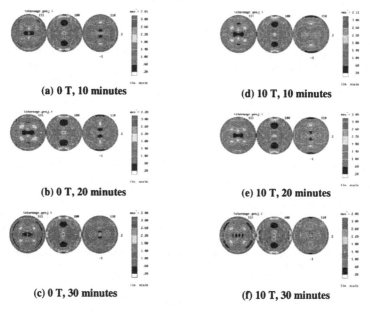

(a) 0 T, 10 minutes   **(d) 10 T, 10 minutes**

(b) 0 T, 20 minutes   **(e) 10 T, 20 minutes**

(c) 0 T, 30 minutes   **(f) 10 T, 30 minutes**

Figure 2: Complete pole figures for the samples processed in a 0 and 10 Tesla magnetic fields at 400°C for 10, 20 and 30 minutes respectively.

116

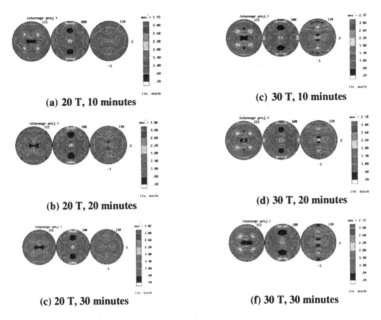

(a) 20 T, 10 minutes

(b) 20 T, 20 minutes

(c) 20 T, 30 minutes

(c) 30 T, 10 minutes

(d) 30 T, 20 minutes

(f) 30 T, 30 minutes

Figure 3: Complete pole figures for the samples processed in a 20 and 30 Tesla magnetic fields at 400°C for 10, 20 and 30 minutes respectively.

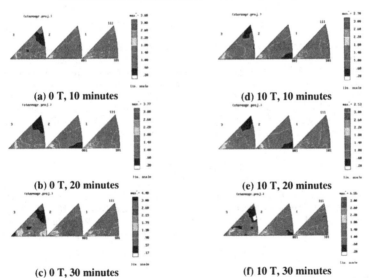

(a) 0 T, 10 minutes

(b) 0 T, 20 minutes

(c) 0 T, 30 minutes

(d) 10 T, 10 minutes

(e) 10 T, 20 minutes

(f) 10 T, 30 minutes

Figure 4: Inverse pole figures for the samples processed in a 0 and 10 Tesla magnetic fields at 400°C for 10, 20 and 30 minutes respectively.

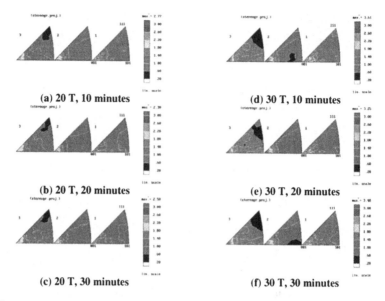

(a) 20 T, 10 minutes	(d) 30 T, 10 minutes
(b) 20 T, 20 minutes	(e) 30 T, 20 minutes
(c) 20 T, 30 minutes	(f) 30 T, 30 minutes

Figure 5: Inverse pole figures for the samples processed in a 20 and 30 Tesla magnetic fields at 400°C for 10, 20 and 30 minutes respectively.

Conclusions

It is concluded from the (111) pole figures that the texture exhibits a close to ideal rolling texture in Figure 1. However, further processing of the rolled alloy through the annealing at 400°C in the magnetic fields and for various periods of time as indicated in Figures 2, 3, 4 and 5 shows evidence of deviation from ideal rolling texture. This is thought to be attributed to successive change of dislocations and the retained texture that have accumulated during rolling. The magnetic field has been found not to enhance the texture with respect to the complete and inverse pole figures. However, it has been reported elsewhere by Adedokun, et al [1], [2], [3] how texture changes occurred with changes in orientation densities for aluminum alloy 6061 heat treated in magnetic fields.

Acknowledgements

The author would like to thank the University of Lagos, Akoka-Yaba, Lagos, Nigeria for the opportunity given me to conduct this research work at the National High Magnetic Field Laboratory in the US. Also, thanks for the funding provided by Dr Reginald Perry through the Minority Doctoral Engineering Fellowship funding by Title III is US Department of Education.

References

1. S. T. Adedokun and S. A. Adedokun (2008) "Magnetic annealing aluminum alloy 6061 at 400°C in different magnetic fields" MS&T Conference. October 5-9, 2008, Pittsburg, Pennsylvania, USA.

2. S. T. Adedokun, Akin Fashanu, Bayo Ogunmola (2008) "Effects of Heat Treating Aluminum alloy 6061 in 8 T and 20 T magnetic fields" *2008* TMS Conference, New Orleans, Louisiana.

3. S. T. Adedokun, J, Schwartz and H. Garmestani (2007) "Texture variation in an heat treated heavily deformed aluminum alloy 6061" Materials Science & Technology 2007 Conference Proceedings, pages 297 – 302.

4. Bacaltchuk, Cristiane Maria Basto (2005) "The effect of magnetic annealing on the texture and microstructure development in silicon steel", The Florida State University PhD Dissertation.

5. S. Bhaumik, X. Molodova, D.A. Molodov and G. Gottstein (2006) "Magnetically enhanced recrystallization in an aluminum alloy" Scripta Materialia 55, pages 995-998.

6. Chen (1998) "Relationship between texture and magnetic properties of non-oriented Si steel" The Third Pacific Rim International Conference on Advanced Materials and Processing (PRICM 3), pp223-228.

7. Cullity, B. D. (2001) "Elements of X-Ray Diffraction" Prentice Hall, New Jersey, 2001.

8. Kocks et al (1998) "Texture and Anisotropy: Preferred Orientations in Polycrystals and their Effect on Material Properties" Cambridge University Press.

9. H. O. Martikainen. and V. K. Lindroos (1981) "Observations on the effect of magnetic field on the recrystallization in ferrite" Scandinavian Journal of Metallurgy, vol. 10, pp. 3.

10. D.A. Molodov, S. Bhaumik, X. Molodova and G. Gottstein (2006) 'Annealing behavior of cold rolled aluminum alloy in a high magnetic field' Scripta Materialia 54, pages 2161-2164.

11. A. D. Sheikh-Ali, D. A. Molodov and H. Garmestani (2002) 'Magnetically induced texture development in zinc alloy sheet' Scripta Materialia 46, pages 857-862.

Aluminum Alloys:
Fabrication, Characterization and Applications II

Materials Characterization

Session Chair

Sooho Kim

Aluminum Alloys: Fabrication, Characterization and Applications II
Edited by: Weimin Yin, Subodh K. Das, and Zhengdong Long
TMS (The Minerals, Metals & Materials Society), 2009

INFLUENCE OF HEAT TREATMENT ON LOW-CYCLE FATIGUE BEHAVIOR OF AN EXTRUDED 6063 ALUMINUM ALLOY

Chunyan Ma[1], Lijia Chen[1,2], Yuxing Tian[1], Xin Che[1], P K Liaw[2]

[1]School of Materials Science and Engineering, Shenyang University of Technology;
111 Shenliao Xi Road, Shenyang Economic and Technology Development Zone;
Shenyang 110178, P R China
[2]Department of Materials Science and Engineering, The University of Tennessee;
Knoxville, TN 37996-2200, USA

Keywords: aluminum alloy, low-cycle fatigue, heat treatment, cyclic stress response, fatigue life

Abstract

Low-cycle fatigue studies were performed under the total strain-amplitude-controlled mode for the extruded 6063 aluminum alloys with different heat-treatment states. The influence of heat treatment on the fatigue behavior of the alloy was determined. The experimental results show that the alloys with different heat-treatment conditions exhibit cyclic hardening, softening and stability. The solution plus aging treatment can increase the fatigue life of the alloy, while the solution treatment leads to a decrease in the fatigue life of the alloy. For the as-extruded 6063 alloy, a single-slope linear relation between the elastic-strain amplitude, or the plastic-strain amplitude, and reversals to failure is observed. However, for the extruded 6063 alloys subjected to both solution and solution plus aging treatments, the single-slope linear relation between the elastic-strain amplitude and reversals to failure is noted while a two-slope linear relation between the plastic-strain amplitude and reversals to failure is noted.

Introduction

Aluminum alloys, due to their high strength to weight ratio, corrosion resistance, good thermal conductivity and excellent workability, are widely employed for the manufacture of parts and components in the transportation and construction industries. Especially they are being increasingly used in vehicle structures for weight reduction and fuel economy improvement. The 6000 series (Al-Mg-Si based) aluminum alloys is one of the most important groups of alloys for the extrusion of plates, rods, tubes and other shapes. These alloys have good formability and corrosion resistance with medium strength and very good surface finish[1]. The 6063 alloy is the most widely used of all extruded aluminum alloys for structural applications[2]. The alloy mainly contains silicon and magnesium in the appropriate proportions to form magnesium silicide, which makes it heat treatable and can reach full properties with the artificial aging. In the 6063 alloy, such the elements as Cu, Mn, Cr, Ti and Zn may be present below the specified maximum limits.

Although extensive studies exist on the precipitation sequence, microstructures and mechanical properties of the 6063 aluminum alloy subjected to various processes and heat treatments[3-7], a systematic study of the effect of heat treatment on the fatigue behavior of this alloy has not been reported. In this study, the low-cycle fatigue behaviors of the extruded 6063 alloys with and without heat treatments are presented. The influence of solution and aging treatments on the fatigue properties of the extruded 6063 alloy has been identified, with particular emphasis on the cyclic stress response and fatigue life behaviors of the alloy.

123

Experimental procedures

The 6063 aluminum alloy casting ingots were homogenized at 520°C for 7 h in a batch furnace and cooled to room temperature using forced air. The homogenized ingots pre-heated at 450°C for about 15 min were extruded in a 630 tons horizontal hydraulic press with a press exit temperature of 465°C and an extrusion ratio of 35. The extruded round bars with a diameter of 12 mm were cooled down to room temperature with forced air. The fatigue specimens with a gauge length of 10 mm and a gauge diameter of 6 mm were machined from the extruded bars, where the specimen axis is parallel to the extrusion direction. The low-stress machining was practiced in fabricating the specimens so as to minimize the mechanical damage on the surface of the specimens. Some fatigue specimens were solution treated at 520°C for 1.5 h and cooled to room temperature with water quench. For a part of the fatigue specimens subjected to solution treatment, the artificial aging treatment was performed at 175°C for 8 h and air cooled to room temperature.

Before the fatigue tests, all specimens were ground manually using 600-grit silicon carbide paper to remove any circumferential scratches and surface machining marks. The fatigue tests were performed on a PLD-50 servohydraulic test machine under total strain-control mode at room temperature. The imposed total strain amplitude was ranged from 0.3% to 1.0%. A triangular waveform was applied, and the cyclic frequency was 0.5 Hz. All fatigue tests, where at least two specimens were used at each given strain amplitude, were run to the failure or separation of the specimens. The number of cycles-to-failure or separation was taken as fatigue life, N_f.

Results and Discussion

Cyclic stress response behavior

Figure 1 illustrates the cyclic stress response curves of the extruded 6063 alloys with different heat-treatment states at the different imposed total strain amplitudes. It can be seen from Figure 1a that over the entire range of total strain amplitudes used in the present investigation, as-extruded 6063 alloy exhibits the cyclic strain hardening till a rapid stress drop due to initiation and growth of fatigue cracks. The degree of cyclic strain hardening is relatively lower in the initial stage of fatigue deformation. However, the degree of cyclic strain hardening gets considerably enhanced in the latter stage of fatigue deformation, which is much more significant at those lower total strain amplitudes. As shown in Figure 1b, the cyclic stress response behavior of the extruded 6063 alloy subjected to solution treatment is similar to that of as-extruded 6063 alloy, that is, the alloy exhibits the cyclic strain hardening at different total strain amplitudes. At the total strain amplitudes ranged from 0.6% to 1.0%, the degree of cyclic strain hardening is almost the same and independent on the imposed total strain amplitude. When the total strain amplitudes of 0.3% and 0.45% are imposed, the degree of cyclic strain hardening is lower at the initial stage of fatigue deformation and then gets significantly increased. It can be observed from Figure 1c that the extruded 6063 alloy subjected to solution plus aging treatment shows a different cyclic stress response from either as-extruded 6063 alloy or the extruded 6063 alloy subjected to solution treatment. At the total strain amplitudes ranged from 0.45% to 1.0%, the alloy exhibits the initial cyclic strain hardening followed by a stable cyclic stress response. At the total strain amplitude of 0.3%, the alloy shows the initial cyclic strain hardening, and then the cyclic stability is reached, while in the latter stage of fatigue deformation, the cyclic softening can be observed.

It can be thought that the occurrence of the cyclic strain hardening is closely associated with the interaction of many mobile dislocations. A mass of dislocations initiate interiorly during cyclic loading and are likely to tangle with each other. Subsequently, the complicated microstructures such as the dislocation tangles and Lomer-Cottrel lock form. These tangles and locks contribute to barriers for the further motion of slip dislocation and the mobility of dislocations decreases, which will result in the cyclic strain hardening. Moreover, the dynamic interactions between mobile dislocations and diffusing solute atoms will occur during fatigue deformation, which is also a paramount factor resulting in the cyclic strain hardening. Solute atoms diffuse towards dislocations that are temporarily arrested. Subsequently, solute atmospheres develop, which restrict the movement of dislocations. For the extruded 6063 alloy subjected to solution plus aging treatment, the cyclic strain softening is mainly related to a continual reduction in the size of the precipitates by shearing during fatigue deformation. When the hardening effect is balanced by the softening effect, the stable cyclic stress response can be attained.

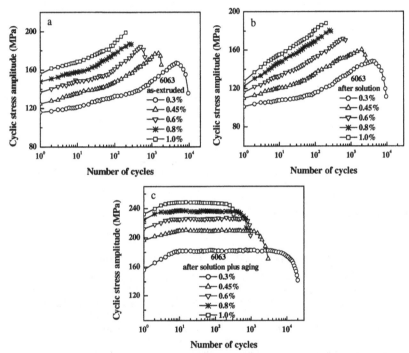

Figure 1. Cyclic stress response curves of extruded 6063 alloys with different treatment states a) as-extruded; b) after solution; c) after solution plus aging

Fatigue life behavior

Figure 2 shows the total strain amplitude versus fatigue life curves for the extruded 6063 alloys with different treatment states. It can be noted that at those total strain amplitudes ranged from 0.6% to 1.0%, the extruded 6063 alloy subjected to solution plus aging treatment gives the

125

longest fatigue life, while the fatigue life of the extruded 6063 alloy subjected to solution treatment is very close to that of as-extruded alloy. At the total strain amplitude of 0.45%, the extruded 6063 alloys after solution and solution plus aging treatments show the longer fatigue life than as-extruded 6063 alloy. At the total strain amplitude, the fatigue life of the extruded 6063 alloy subjected to solution plus aging treatment remains the longest, while there is almost no difference in fatigue lives of the extruded 6063 alloy subjected to solution treatment and as-extruded alloy. The above-mentioned fact implies that the solution plus aging treatment can effectively enhance the fatigue resistance of the extruded 6063 alloy, while the solution treatment has little influence on the fatigue life of the extruded 6063 alloy.

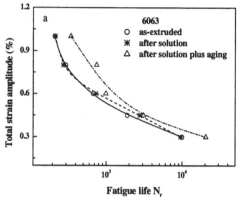

Figure 2 Total strain amplitude versus fatigue life curves
for extruded 6063 alloys with different treatment states

For the total strain-controlled low cycle fatigue test, the relationship between the imposed total strain amplitude ($\Delta \varepsilon_t / 2$) and reversals to failure ($2N_f$) can be expressed as

$$\Delta \varepsilon_t / 2 = \Delta \varepsilon_p / 2 + \Delta \varepsilon_e / 2 = \varepsilon_f'(2N_f)^{-c} + \sigma_f'(2N_f)^{-b} \tag{1}$$

where $\Delta \varepsilon_p / 2$ and $\Delta \varepsilon_e / 2$ are plastic and elastic-strain amplitudes respectively, ε_f' is the fatigue ductility coefficient, c is the fatigue ductility exponent, σ_f' is the fatigue strength coefficient and b is the fatigue strength exponent.

Figure 3 shows the curves of three strain amplitudes versus reversals to failure for the extruded 6063 alloy with different treatment states. It is obvious that the relation between elastic-strain amplitude with reversals to failure shows a monotonic linear behavior for the extruded 6063 alloys with different treatment states. For the as-extruded 6063 alloy, a single-slope linear relation between plastic-strain amplitude and reversals to failure is observed. However, for the extruded 6063 alloys subjected to both solution and solution plus aging treatments, a two-slope linear relation between the plastic-strain amplitude and reversals to failure is noted. Through the linear regression analysis, the corresponding strain fatigue parameters for the extruded 6063 alloys with different treatment states can be determined and are listed in Table I.

Cyclic stress-strain behavior

The cyclic stress-strain curves of the extruded 6063 alloys with different treatment states are shown

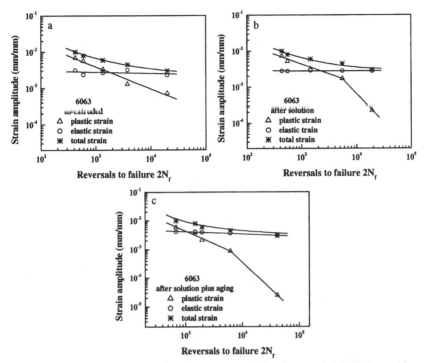

Figure 3 Strain amplitude as a function of reversals to failure for extruded 6063 alloys with different treatment states a) as-extruded; b) after solution; c) after solution plus aging

Table I. Strain Fatigue Parameters of Extruded 6063 Alloys with Different Processing States

States of Alloy	σ_f' (MPa)	b	ε_f' (%)	c	K' (MPa)	n'
As-extruded	243.95	-0.0360	26.34	-0.6099	235.04	0.0475
After Solution	194.82	-0.0017	17.17	-0.5399	227.50	0.0531
After Solution Plus Aging	512.22	-0.0847	220.14	-0.8974	333.88	0.0616

in Figure 4. It is obvious that the cyclic stress amplitude and plastic strain amplitude exhibit a linear relationship. Usually, the relationship between the cyclic stress amplitude and plastic strain amplitude can be expressed by following power law, that is

$$\Delta\sigma/2 = K'(\Delta\varepsilon_p/2)^{n'} \qquad (2)$$

where K' is the cyclic strength coefficient, and n' is the cyclic strain hardening exponent. Based on the data in Figure 4, the strain fatigue parameters K' and n' of the extruded 6063 alloys with different treatment states can be determined by the linear regression analysis, and the values of K' and n' are also listed in Table I.

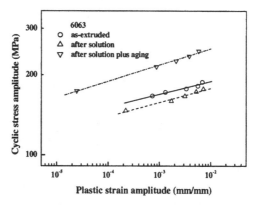

Figure 4 Cyclic stress-strain curves for extruded 6063 alloys with different treatment states

Conclusions

1. The extruded 6063 alloy subjected to solution treatment and as-extruded 6063 alloy, and exhibits the cyclic strain hardening during whole fatigue deformation, while the extruded 6063 alloy subjected to solution plus aging treatment mainly shows the initial cyclic strain hardening followed by a stable cyclic stress response.
2. The solution plus aging treatment can effectively increase the fatigue life of the extruded 6063 alloy, while the solution treatment has no significant influence on the fatigue life of the extruded 6063 alloy.
3. For the as-extruded 6063 alloy, a single-slope linear relation between the plastic-strain amplitude and reversals to failure is observed. However, for the extruded 6063 alloys subjected to both solution and solution plus aging treatments, a two-slope linear relation between the plastic-strain amplitude and reversals to failure is noted.

References

1. J.E. Hatch ed., *Aluminum: Properties and Physical Metallurgy* (Metals Park, OH: ASM, 1984), 64.
2. J.R. Davis ed., *Aluminium and aluminium alloys* (Materials Park, OH: ASM International, 1993), 271.
3. A. Piñeiro-Jiménez et al., "Tensile and fatigue properties of 6063-T6 aluminium alloy coated with electroless Ni–P deposit," *Materials Science and Technology*, 23 (2007), 253-263.
4. K.B.S. Couto et al., "Effects of homogenisation treatment on microstructure and hot ductility of aluminium alloy 6063," *Materials Science and Technology*, 21 (2005), 263-268.
5. A.A. Luo, R.C. Kubic, and J.M. Tartaglia, "Microstructure and fatigue properties of hydroformed aluminum alloys 6063 and 5754," *Metallurgical and Materials Transactions A*, 34A (2003), 2549-2557.
6. A. Munitz, C. Cotler, and M. Talianker, "Aging impact on mechanical properties and microstructure of Al-6063," *Journal of Materials Science*, 35 (2000), 2529-2538.
7. Y.S. SATO et al., "Precipitation sequence in friction stir weld of 6063 aluminum during aging," *Metallurgical and Materials Transactions A*, 30A (1999), 3125-3130.

Aluminum Alloys: Fabrication, Characterization and Applications II
Edited by: Weimin Yin, Subodh K. Das, and Zhengdong Long
TMS (The Minerals, Metals & Materials Society), 2009

Precipitation Under Cyclic Strain in Solution-Treated Al-4wt%Cu I: Mechanical Behavior

Adam Farrow[1] and Campbell Laird[2]

[1]Los Alamos National Laboratory; MS E574 Los Alamos, NM 87544, afarrow@lanl.gov
[2]University of Pennsylvania; 3231 Walnut St. Philadelphia, PA 19104

Keywords: Aluminum, Cyclic Strain, Diffusion, Precipitation

Abstract

Solution-treated Al-4wt%Cu was strain-cycled at ambient temperature and above, and the precipitation and deformation behaviors investigated by TEM. Anomalously rapid growth of precipitates appears to have been facilitated by a vacancy super-saturation generated by cyclic strain and the presence of a continually refreshed dislocation density to provide heterogeneous nucleation sites. Texture effects as characterized by Orientation Imaging Microscopy appear to be responsible for latent hardening in specimens tested at room temperature, with increasing temperatures leading to a gradual hardening throughout life due to precipitation. Specimens exhibiting rapid precipitation hardening appear to show a greater effect of texture due to the increased stress required to cut precipitates in specimens machined from rolled plate at an angle corresponding to a lower average Schmid factor. The accelerated formation of grain boundary precipitates appears to be partially responsible for rapid inter-granular fatigue failure at elevated temperatures, producing fatigue striations and ductile dimples coexistent on the fracture surface.

Introduction

Based upon past studies [1-3] significant changes in the precipitation behavior of Al-4wt%Cu seemed likely to occur under cyclic plasticity. Elevated point-defect concentrations due to deformation were expected to accelerate diffusion kinetics [2], and the presence of pinned dislocations to act as nucleation sites were expected to accelerate precipitate nucleation rates [4]. The role of glissile dislocations in precipitation under fatigue is less clear. Past studies have shown dissolution of Θ' [5], and Θ'' [6] precipitates under fatigue, and have postulated that precipitates may be softened by the introduction of anti-phase domain boundaries under deformation [7]. A transient increase in precipitate thickening rates has also been observed following monotonic deformation [8]. The combined effects of these phenomena upon cyclic-strain assisted precipitate growth from the solution state have never been investigated. Solution-treated material was desired for this study in order to attempt to understand the effects of cyclic strain. Material aged to contain fine precipitates has a tendency towards strain localization [7], leaving a great deal of uncertainty as to the strains experienced by any area inspected by TEM. Material aged to contain coarse precipitates would contain little remaining copper in the aluminum matrix, and thus a very low driving force for any precipitation during testing. Solution-treated material presented a clean slate with respect to precipitation, allowing for simple comparison of precipitation at different testing conditions, a high driving force for precipitation, and a tendency to strain more homogenously than material containing fine precipitates. Additionally, the role of texture hardening during precipitation hardening was investigated in an effort to deconvolute precipitation hardening from texture hardening.

Experimental Details

Two 12.7 kg (28 lb) ingots of 99.99% aluminum with 4 weight % copper added were cast by Alcoa, homogenized at 540°C (1000°F) for twelve hours, and hot rolled to 3/8" plate at an entry temperature of 400°C (750°F). Material was received in this condition and further cold rolled to 7 mm (0.275"), following which standard "dog-bone" specimens were machined at 0° and 45° from the rolling direction. Specimens were then dry ground by to a 600 grit finish and wet ground to a 1200 grit finish over a pane of glass. They were then annealed at 540°C in molten salt for ten minutes and quenched into ice water immediately prior to mechanical testing. Orientation Imaging Microscopy (OIM) was performed on a FEI XL30 SEM equipped with TSL OIM 4.0 software. This showed a mixture of cube and fiber textures, with the cube texture predominant. Due to the very large areas sampled in order to gain statistical significance with a grain size of 350 µm, it was necessary to condense the data from multiple low-magnification scans to allow for an overall picture of the texture of the material. Pole figures are shown in Figure 1.

Following the generation of pole figures via the OIM software package, the raw data were further processed to generate a locus of the average Schmid factors (the purely geometric resolution of axial stress onto shear stress for a given slip system) of the material. This was achieved by transforming the Euler angles generated by the OIM system into Miller indices, and calculating the resolved stress on each slip system for any angle of application of unit stress in the plane of the plate between 0° from the rolling direction and 45° from the rolling direction, and then averaging the results for each angle across all the data points gathered.

Specimens machined with their stress axes parallel to the rolling direction of the parent plate were tested at plastic strain amplitudes of +/- 0.001 at 1 Hz, +/- 0.0025 at 0.4 Hz and +/-0.005 at 0.2 Hz and at temperatures of 25°C, 100°C, 175°C and 200°C in a forced hot air furnace on an Instron 1331 servo-hydraulic test system. The furnace was pre-heated and air flow started at the same time as the actuator, in order to ensure that the specimens began the test in a solution-treated condition.

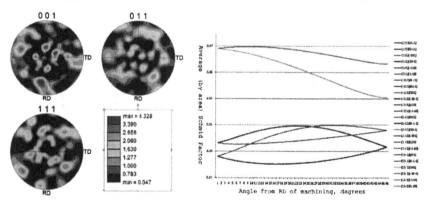

Figure 1. Pole figures for the material as tested, and average Schmid factors for each slip system calculated from the OIM data. All slip systems have been plotted as a mathematical check despite the overlay between many systems.

The accumulated plastic strain at failure for the specimen tested at ε_p=+/- 0.005 was used as a test-stop criterion for specimens tested at lower strain amplitudes. Thus each specimen

130

accumulated equal plastic strain during the test, and each test duration was the same for a given temperature due to the inverse relationship between frequency and strain amplitude.

Specimens machined from their parent plate with their stress axes at 45° from the rolling direction were tested to failure at ε_p=+/- 0.001 at 1 Hz and ε_p=+/-0.0025 at 0.4 Hz.

Results and Discussion

<u>Cyclic Strain Behavior in the Rolling Direction</u>

Cyclic hardening curves for specimens machined with their stress axes parallel to the rolling direction of the parent plate are presented in Figure 2. Of note in these figures is the softening and re-hardening of specimens tested at 175°C and 200°C. Due to the test procedure of starting the forced hot-air furnace at the same time as the actuator, all specimens tested at elevated temperature undergo a short ramp to temperature during the initial strain cycles of mechanical testing. The early softening is caused by recovery of cyclic hardening accumulated at lower temperatures, while the subsequent hardening is an effect of precipitation hardening occurring during testing. At 100°C, although recovery is not apparent, precipitation hardening can be seen in the increasing stress through the plateau region as compared to tests run at 25°C.

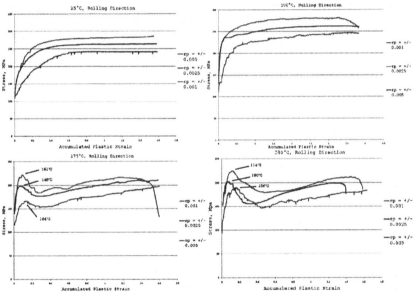

Figure 2. Cyclic hardening curves for specimens tested parallel to the rolling direction of the parent plate. Note that the abscissa on each plot is accumulated plastic strain, and not cycles, to allow for comparison between different strain amplitudes.

<u>Cyclic Strain Behavior at 45° from the Rolling Direction</u>

Cyclic hardening curves for specimens machined at 45° from the rolling direction and tested at ε_p=+/-0.001 at 1 Hz and ε_p=+/-0.0025 at 0.4 Hz are shown in Figure 2. Specimens oriented at 45° from the rolling direction show higher stresses for a given strain amplitude, and tend to fail at lower accumulated strains than specimens machined at 0° from the rolling direction. This effect is less pronounced at ε_p=+/-0.0025, 0.4 Hz than at ε_p=+/-0.001, 1 Hz. It appears that the difference between Schmid factors for secondary slip at 0° and 45° from the rolling direction roughly explains the difference in stresses for specimens strained at ε_p=+/-0.001, 1 Hz, while the difference in Schmid factors for primary slip aligns more closely with specimens strained at ε_p=+/-0.0025, 0.4 Hz. In light of this it seems likely, given the peak prior to the plateau stresses observed in tests performed at 25°C, that latent hardening is a significant factor in these observed texture effects, and is serving to block the activation of a second slip system at low strain amplitudes. The diminishing effect of orientation at ε_p=+/-0.0025, 0.4 Hz suggests that this may be the case. The effect of texture also becomes less pronounced at higher temperatures, suggesting that thermal activation of cross-slip or thermal climb may play a role in reducing the effects of texture under these test conditions. Additional confirmation for this hypothesis comes from the temperatures at which recovery becomes apparent in specimens tested at 175°C and 200°C. During the ramp to temperature early in specimen life, specimens machined from the parent plate at 45° to the rolling direction appear to recover at lower temperatures than specimens oriented in the rolling direction, suggesting a higher dislocation density for a given strain amplitude in specimens oriented at 45° from the parent plate. This effect also disappears at ε_p=+/-0.0025, 0.4 Hz. It appears that at high temperatures, where significant precipitation occurs, texture may change the effectiveness of precipitation hardening. At 200°C, it is apparent that texture hardening complements precipitation hardening, as specimens machined from the parent plate at 45° from the rolling direction appear to harden at a faster rate than those machined parallel to the rolling direction. This effect appears more subtly at 175°C, and is more apparent in the specimen strained at ε_p=+/-0.0025, 0.4 Hz. The likely explanation for this is that greater axial stresses are required for a sufficient force on a dislocation to cut a precipitate.

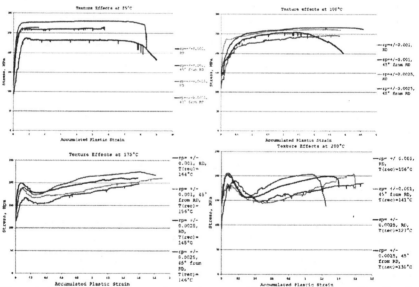

Figure 3. Cyclic hardening curves for specimens machined at 45° from the rolling direction of the parent plate. Temperatures corresponding to the early peak are listed in the legend for specimens tested at 175°C and 200°C.

132

Time-Dependent Effects at 200C

Several additional tests were performed at 200°C, in order to determine the effects of frequency and holds on the observed precipitation and mechanical behavior of the material. Frequency was reduced to 100 cycles/hour (0.027 Hz) for tests performed at ε_p=+/-0.005 and the test results compared against those generated at 0.2 Hz. Precipitation hardening is shown to dominate the hardening behavior of the material at 200°C via comparison of the results at 0.2 Hz and 0.027 Hz. By plotting results against time rather than accumulated plastic strain, and comparing the slopes of the cyclic hardening curves, the effect of strain hardening is shown to be minor relative to precipitation hardening. The strain accumulated by the 0.027 Hz test shown in Figure 4 would be insufficient to saturate the strain-hardening behavior at room temperature, and yet the slopes of the two curves are substantially similar. Another view of the same conclusion can be drawn from a 0.027Hz test incorporating strain holds. Also shown in Figure 3, the recovery behavior is dominated by precipitation hardening upon resumption of strain-cycling. That is: although the load drops during a hold, the cycle immediately following the hold reveals a stronger material, rather than a softened one.

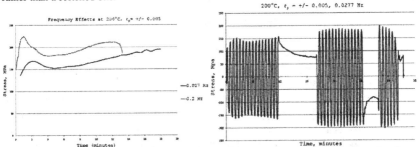

Figure 4. Frequency effects and recovery behavior during holds at 200°C. Note the change in the abscissa to time from accumulated plastic strain to allow comparison of hardening effects. Samples sectioned from parent plate parallel to the rolling direction.

Fracture Behavior.

At 25°C and 100°C, failure proceeded as expected, via nucleation and growth of a transgranular fatigue crack. At 175°C and 200°C, the increased diffusion of solute under cyclic strain promoted a ductile-intergranular fracture mode featuring both ductile dimples and fatigue striations on the fracture surface. It is believed that the presence of a narrow precipitate free zone accompanied by an accelerated growth of grain boundary precipitates is responsible for this fracture mode. Previous authors [3] have reported on the role of Θ' precipitation in the bulk in promoting grain boundary fracture, due to increased prevalence in multiple slip allowing for greater cooperative deformation across boundaries. The observed fracture mode persists at all strain amplitudes sampled at 175°C and 200°C, and persists in specimens of different textures. Initially, dynamic embrittlement was considered a possible cause of this fracture behavior, but referencing the strain-hold test shown in figure 3, it seems unlikely that hold crack plays a role, given the smooth stress drops and the absence of a faster stress drop in tension than compression. In specimens with a growing crack stopped prior to final failure, cooled to room temperature, and broken in monotonic tension, the same grain boundary failure mode persists, with ductile dimples apparent on the grain boundaries, and no striations observed in the region of monotonic failure. Examples of the fracture surface generated by this type of crack propagation can be seen in Figure 4.

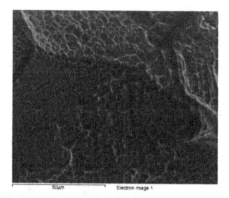

Figure 5. Scanning electron micrograph, showing ductile dimples and fatigue striations on the fracture surface of a specimen machined at $45°$ from the parent plate and tested to failure (800 cycles) at $\varepsilon_p=+/-0.001$, 1 Hz, 200°C.

Conclusions

1. Precipitation hardening dominates the high-temperature cyclic strain behavior of Al-4wt%Cu from the solution state, due to accelerated formation of Θ' precipitates.

2. Enhanced precipitation kinetics described in a companion paper promote the formation of ductile dimples in the precipitate free zone, and precipitation hardens the surrounding material, enhancing strain localization at grain boundaries at 175°C and 200°C leading to intergranular failure by a hybrid stage II fatigue crack and void coalescence mechanism.

3. Texture hardening appears to diminish with both increasing temperature and increasing strain amplitude, but is accentuated by precipitation.

References

1. Ramanathan Sankaran. *Misfit Dislocation Structures, Growth Kinetics, and Morphology of Platelike Precipitates in Al-0.2%Au and Al-4%Cu Alloys*, PhD thesis, University of Pennsylvania (1973).
2. C. Laird, VJ Langelo, M Hollrah, NC Yang and R De Le Veaux, "The Cyclic Stress-Strain Response of Precipitation Hardened Al-15wt%Ag Alloy." *Mat. Sci. Eng.* **A32,** 137-160 (1978).
3. HD Chandler and JV Bee. "Cyclic Strain Induced Precipitation in a Solution-Treated Aluminum Alloy." *Acta Metall.* **35**:10, 2503-2510, (1987).
4. J Cahn. "Nucleation on Dislocations." *Acta Metall.* **5**, 169-172, (1957).
5. Shrikant Bhat. *High Temperature Cyclic Deformation in Nickel, TD-Nickel, and Al-4%Cu Alloy aged to contain Θ' and Θ''*. PhD Thesis, University of Pennsylvania (1978).
6. C Lea, SJ Brett, and RD Doherty. "Solute Depletion at Fatigue Facets in Precipitation Strengthened Aluminum Alloys – Auger Electron Spectroscopy." *Scripta Met.* **13**, 45-50, (1979).
7. Carmen Calabrese. *Cyclic Response and Fatigue Life Prediction of Two Phase Alloys*. PhD Thesis. University of Pennsylvania (1972).
8. R Sankaran and C Laird. "The Role of Intruder Dislocations in Modifying the Misfit Dislocation Structures and Growth Kinetics of Precipitates." *Met Trans.* **5**, 1794-1803, (1974).

Aluminum Alloys: Fabrication, Characterization and Applications II
Edited by: Weimin Yin, Subodh K. Das, and Zhengdong Long
TMS (The Minerals, Metals & Materials Society), 2009

Precipitation Under Cyclic Strain in Solution-Treated Al-4wt%Cu II: Precipitation Behavior

Adam Farrow[1] and Campbell Laird[2]

[1]Los Alamos National Laboratory; MS E574 Los Alamos, NM 87544, afarrow@lanl.gov
[2]University of Pennsylvania; 3231 Walnut St. Philadelphia, PA 19104

Keywords: Aluminum, Cyclic Strain, Diffusion, Precipitation

Abstract

Solution-treated Al-4wt%Cu was strain-cycled at ambient temperature and above, and the precipitation behavior investigated by Transmission Electron Microscopy (TEM). In the temperature range 100°C to 200°C, precipitation of Θ'' appears to have been suppressed, and precipitation of theta-prime promoted. Anomalously rapid growth of precipitates appears to have been facilitated by a vacancy super-saturation generated by cyclic strain, with a diminishing effect observed at higher temperatures due to the recovery of non-equilibrium vacancy concentrations. The Θ' precipitates generated under cyclic strain are considerably smaller and more finely dispersed than those typically produced via quench-aging due to their heterogeneous nucleation on dislocations, and possess a low aspect ratio and rounded edges of the broad faces, due to the introduction of ledges into the growing precipitates by dislocation cutting. Frequency effects indicate that dislocation motion, rather than extremely small precipitate size, is responsible for the reduction in aspect ratio.

Introduction

Accelerated hardening kinetics under cyclic strain reported in a companion paper are explored via a line of transmission electron microscope studies to determine the nucleation and growth behaviors of precipitates formed under cyclic strain. Measurements taken from TEM images are used to estimate precipitate growth rates, enhanced diffusivities under cyclic strain, elevated vacancy concentrations, and characterize unusual precipitate aspect ratios.

Experimental Details

Following processing described in a companion paper, dog-bone specimens of Al-4wt%Cu machined parallel to the rolling direction of their parent plate were annealed in molten salt at 540°C for 10 minutes, and were quenched into ice water. Immediately following quenching, they were strain-cycled at $\varepsilon_p=+/- 0.001$, 1 Hz, $\varepsilon_p=+/- 0.0025$, 0.4 Hz, and at $\varepsilon_p=+/- 0.005$, 0.2 Hz, at temperatures of 25°C, 100°C, 175°C and 200°C. Tests performed at $\varepsilon_p=+/- 0.001$, 1 Hz, and $\varepsilon_p=+/- 0.0025$, 0.4 Hz were interrupted at equal accumulated strain to the strain accumulated at failure in tests run at $\varepsilon_p=+/- 0.005$, 0.2 Hz at each temperature. Thus, each specimen accumulated equal plastic strain and spent equal time at temperature due to the inverse variation of test frequency with strain amplitude. An additional test was run at 200°C, $\varepsilon_p=+/- 0.005$, 0.027 Hz (100 cycles/Hr) to assess the effects of frequency. Following testing, samples were rapidly cooled by raising the pressure of air fed to the furnace and turning off the heating elements.

TEM foils of each specimen were then prepared by cutting transverse to the stress axis with a jeweler's saw to 0.5mm sections, chemical thinning in 25g NaOH in 100 mL H_2O at 25°C, and electropolishing by the window method in 25mL fuming nitric acid in 75 mL methanol at 8V, -30°C. 3mm disks were then punched from the thinned foils and mounted in a double-tilt holder in a JEOL JEM 2010 transmission electron microscope. Bright-field imaging and selected area diffraction were used to characterize precipitates formed during mechanical testing. Following imaging, a representative sampling of at least 25 precipitates from each specimen displaying precipitation were measured in both length and thickness, and these results were tested against the Zener-Hillert equation for plate lengthening to estimate the diffusivity during testing of each specimen.

Results and Discussion

At 25°C no precipitation was observed, likely due to the short test duration (800 seconds) failing to provide sufficient time for nucleation of precipitates, even given the presence of plentiful dislocations to act as nucleation sites. The dislocation structure formed during these tests appeared largely similar to that observed in pure aluminum [1], consisting of loose tangles of dislocations with small prismatic loops.

At 100°C, fine nucleation of Θ' precipitates was observed in strained specimens, primarily nucleated on dislocations. An example of the structure thus generated can be seen in Figure 1. Precipitates can be seen preferentially nucleating on dislocations, and are confirmed via Selected Area Diffraction (SAD) as Θ', despite their small size, and the relatively short testing time (32 minutes) at 100°C, where Θ" precipitates might be expected. The apparent absence of Θ" implies that not only is Θ' preferentially nucleating, but that cyclic strain inhibits the growth of Θ", as only an extremely small volume fraction of Θ' was observed, suggesting that competitive growth alone cannot account for the absence of Θ".

At 175°C, control specimens annealed on the testframe in the absence of cyclic strain show Θ" precipitates, as expected, but in all cyclic strain tests performed, Θ' precipitates are formed to the apparent exclusion of Θ'. Very little residual dislocation density is apparent, as recovery during cyclic straining has largely annealed out dislocations.

At 200°C, Θ" precipitates are formed in unstrained specimens, as shown in Figure 2. In strained specimens, Θ' rapidly forms, and appears to be sheared by dislocation shuttling under cyclic strain, following a mechanism originally proposed by Bhat [2]. This mechanism was originally proposed to explain precipitate dissolution in larger Θ' precipitates, and the nucleation of Θ on Θ' precipitates under cyclic strain. Slower growth of precipitates was observed in the test cycled at 0.027 Hz, suggesting that the rate of accumulation of plastic strain as well as the plastic strain amplitude helps to control the diffusivity under cyclic strain.

On the basis of the measured average lengths and thicknesses of precipitates for each test, diffusivities were calculated from the Zener-Hillert equation for plate lengthening (Eq. 1).

$$G = \frac{D(\chi\alpha - \chi\alpha\gamma)}{4r(\chi p - \chi\alpha\gamma)} \tag{1}$$

Where G = the precipitate lengthening rate, D = the diffusion coefficient, r = half the precipitate thickness, and $\chi\alpha$, χp, and $\chi\alpha\gamma$ are the solute concentrations in the matrix, precipitate, and at the matrix/precipitate interface, respectively. Given the diffusivities thus calculated and taking the free energy of formation for a vacancy in Al as 0.63 eV [3], the free energy of vacancy motion can be calculated from previous authors' [4-8] data as 0.45 eV, via Eq. 2.

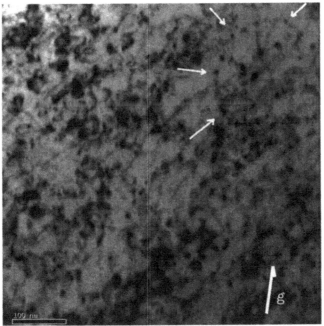

Figure 1. Structure generated at 100°C εp = +/- 0.005, 0.2Hz, 385 cycles (32 minutes) showing precipitation on bowed dislocation (arrowed). g=111, zone axis=[112].

$$D = \frac{1}{6}z\alpha^2 v e^{-\left(\frac{\Delta E_F}{kT}\right)}e^{-\left(\frac{\Delta E_m}{kT}\right)} \tag{2}$$

Where D = diffusivity, z = coordination number, α = inter-atomic distance, v = vibrational frequency of the lattice (taken here as 10^{13}). If the diffusivities calculated in this study are used with the free energy of

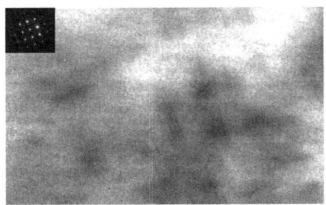

Figure 2. Θ" precipitates formed in an unstrained control specimen annealed at 200°C for 800 seconds. Field of view = 97nm.

137

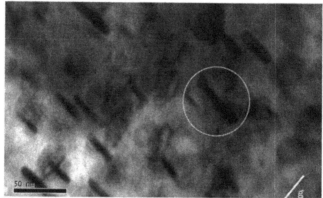

Figure 3. Transmission electron micrograph, showing twisting of Θ' precipitate following repeated shearing. g=(111) of a [110] zone axis. Specimen machined in the rolling direction, tested at εp = +/- 0.0025 to failure at 280 cycles, 0.4 Hz (700 seconds).

motion of 0.45eV, vacancy concentrations can then be approximated for each specimen. The results of this calculation are shown in Figure 5.

Of note is that above recovery temperatures, the vacancy super-saturation begins to drop, as greater mobility allows vacancies to find sinks faster. Thus for a given strain amplitude, a higher vacancy concentration will be present at lower temperatures, since vacancy generation by jog dragging or partial dislocation annihilation will allow for athermal generation of a vacancy super-saturation, while the ability of the vacancies to find sinks is reduced. One obvious anomaly generated by this experimental approach can be seen in the unstrained specimen annealed at 100°C. Due to the annealing above the solvus and quench immediately preceding this intermediate anneal, it seems likely that this super-saturation is residual from the quench. The concentration of thermal vacancies during the initial anneal above the solvus is calculated as $1.25*10^{-4}$, and given the rapid quench and immediate intermediate anneal, it is not unreasonable to believe that this residual vacancy concentration may have accelerated the diffusion kinetics in the specimen sufficiently to accelerate precipitate growth rates to a level corresponding to the steady-state concentration of $6*10^{-7}$ calculated.

A further observation pertains to the aspect ratios of precipitates measured in this study, shown in Figure 6. Given precipitate twisting as evidence of precipitate shearing as proposed by Bhat [2], ledges should be created at precipitate surfaces , allowing for more rapid thickening of precipitates during cyclic strain. Aaronson et. al. [9] calculated energies for Θ':α interfaces as 40 ergs/cm² along the broad faces of plate-like precipitates, and as 445 ergs/cm² along the edges. Thus the equilibrium aspect ratio of precipitates should be expected to be somewhere around 11. Previous authors, (e.g. [10]) have typically measured aspect ratios on the order of 40 for Θ', suggesting that ledge nucleation limits the thickening of plates, as they tend to form at greater than equilibrium aspect ratios. In this study, the continual cutting of growing precipitates leads to an apparent reduction is aspect ratios, and an expected temperature and frequency sensitivity. Since the diffusion rate increases with increasing temperature, and the strain frequencies sampled remain the same across all temperatures, we should expect to see larger aspect ratios at higher temperatures, which appears to be case. Precipitates formed at 100°C show a markedly lower aspect ratio than those formed at higher temperatures. In addition, the specimen strained at ε_p=+/- 0.005, 100 cycles/hr (0.027Hz) shows a much greater aspect ratio than those formed at

138

ε_p=+/- 0.005, 0.2 Hz. Sankaran [11] found a transient increase in Θ' thickening rates following monotonic deformation, but attributed it to a transient increase in diffusivity due to the accelerated diffusion kinetics caused by deformation.

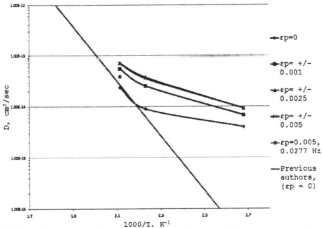

Figure 4. Diffusivities measured in this study as a function of inverse temperature and plastic strain amplitude. Diffusivities for ε_p=+/- 0.0025, 0.4 Hz and ε_p=+/- 0.005, 0.2 Hz overlay. The solid line shows an average of previous authors' interdiffusivities generated by Sankaran [10].

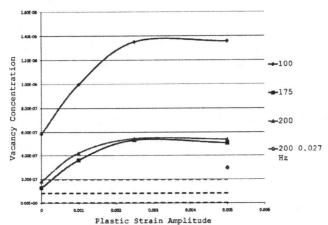

Figure 5. Vacancy concentrations calculated for this study. The dotted lines show the calculated equilibrium thermal vacancy concentrations.

139

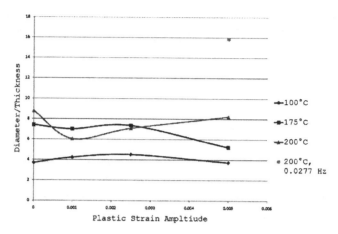

Figure 6. Average aspect ratios of precipitates measured in this study.

Conclusions

1. Cyclic strain effectively suppresses the formation of Θ" and promotes formation of Θ'
2. Diffusion kinetics are appreciably accelerated during cyclic strain at low temperatures, but increased vacancy mobility limits the observed super-saturation at recovery temperatures.
3. The suppression of Θ" precipitates allows for the creation of smaller Θ' precipitates than are observed in conventional quench-aging, as Θ' tends nucleate on dislocations under cyclic strain, rather than on Θ" precipitates, as observed in quench-aging. This allows for smaller nuclei, and nucleation outside of already Cu-enriched precipitates.
4. The aspect ratios of precipitates formed under cyclic strain are significantly reduced via increased thickening rates as ledges are nucleated athermally via dislocation cutting.

References

1. AB Mitchell and DG Teer. "Dislocation Structures in Aluminum Crystals Fatigued in Different Directions." *Metals Science Journal*. **3**, 183-189 (1969).
2. Shrikant Bhat. *High Temperature Cyclic Deformation in Nickel, TD-Nickel, and Al-4%Cu Alloy aged to contain Θ' and Θ"*. PhD Thesis, University of Pennsylvania (1978).
3. KG Lynn and PJ Schultz. *Applied Physics A*. **37**:1, 31-36, (1985).
4. JB Murphy, *Acta Met*. **9**, 563 (1961).
5. MS Anand, SP Muraka and RP Agarwala. *J appl. Phys*. **36**, 3860 (1965).
6. M Beyeler, M Maurice, and R Seguin. *Mem. Sci. Rev. Metall*. **67**, 295 (1970).
7. NL Peterson and SJ Rothman *Phys Rev* **B1**:8, 3264 (1970).
8. S Fujiwara and K Hirano *Trans Jap. Inst. Metals*. **12**, 438 (1971).
9. HI Aaronson, JB Clark, and C Laird. "Interfacial Energy of Dislocation and Coherent Interphase Boundaries." *Metals Sci. J*. **2**, 155-158. (1968).
10. Ramanathan Sankaran. *Misfit Dislocation Structures, Growth Kinetics, and Morphology of Platelike Precipitates in Al-0.2%Au and Al-4%Cu Alloys*, PhD thesis, University of Pennsylvania (1973).
11. R Sankaran and C Laird. "The Role of Intruder Dislocations in Modifying the Misfit Dislocation Structures and Growth Kinetics of Precipitates." *Met Trans*. **5**, 1794-1803, (1974).

Aluminum Alloys: Fabrication, Characterization and Applications II
Edited by: Weimin Yin, Subodh K. Das, and Zhengdong Long
TMS (The Minerals, Metals & Materials Society), 2009

PRECIPITATION AND COARSENING KINETICS AND EFFECT OF MECHANICAL PROPERTY OF Al3Sc IN Al-Sc ALLOY

Qian Wen, Feng Ye*, Meiting Huang, Ji Wang, Junpin Lin

State Key Laboratory for Advanced Metals and Materials, University of Science and Technology
Beijing, Beijing 100083, PR China

Keywords: Al$_3$Sc, Precipitation and coarsening, Electrical conductivity, Mechanical property

Abstract

The present research aims to determine the precipitation and coarsening processes of Al$_3$Sc phase in Al-6Mg-0.2Sc alloy. The electrical resistivity changes are recorded during the isothermal annealing, which reflects the different stages of the nucleation and growth processes of the Al$_3$Sc precipitation. The experimental results show that the Sc addition can improve the mechanical property, and increase the recrystallization temperature of Al-Mg alloys.

Introduction

Due to the high strength, well plasticity and ductility, good corrosion resistance, and especially the excellent weldability, the Al-Sc alloy, following Al-Li alloy, is used as new-style structural material for aeronautical, space and marine applications [1]. Various properties of Al alloys can be improved in many ways by the addition of Sc because of the formation of nanoscale coherent Al$_3$Sc precipitates. The fine dispersed Al$_3$Sc particles can act as grain refiners during solidification, tightly pin up the grain boundaries and the dislocations, inhibit the recrystallization, and match with the matrix during homogenization and solution. This is strengthening mechanism related to the solubility of Sc in Al alloys [2-4].

Al-Mg-Sc alloys belong to a class of advanced structural materials that have been developed in recent years with the purpose of improving some of the service properties of non-heat-treatable Al-Mg alloys [5-7]. The addition of a small amount of Sc to Al-Mg alloys causes a significant increase in the strength of the alloys, due to the existence of coherent, finely dispersed L1$_2$ Al$_3$Sc precipitates [8]. Magnesium is almost insoluble in Al$_3$Sc so that the precipitation kinetics and the hardening effect of Al$_3$Sc precipitates resembles that in binary Al-Sc alloys where the strengthening effect of Al$_3$Sc has been experimentally characterized in a number of studies, see Refs [9-14]. The addition of scandium to Al-Mg significantly increases the hardness and yield stress during aging [15], and can also exhibit good corrosion resistance [5,6] and superplasticity [16]. The present paper aims to investigate the precipitation and coarsening processes of the Al$_3$Sc particles in an Al-Mg-Sc alloy and confirm the excellent mechanical properties of Al-Mg-Sc alloy during the annealing.

Experimental

The tested alloys, Al–6Mg and Al–6Mg–0.2Sc (in weight percent), are prepared in a crucible furnace by remelting commercially pure aluminum (99.5%), Magnesium (99.7%), and Al-3Sc master alloy. The composition is determined by electronic probe micro-analysis (EPMA) on

---

* Corresponding author, email: yefeng@skl.ustb.edu.cn

JEOL JXA-8100 as Al–6.052Mg–0.194Sc–0.336Mn–0.100Fe (wt%). Both Al–6Mg and Al–6Mg–0.2Sc alloys were cold rolled into plates with thickness of ~3 mm. These specimens were solution annealed at 873 K and subjected to aging at 573K, 623K, 673K and 723K and water quenched. Scanning electron microscopy (SEM) observations were carried out on Carl Zeiss SUPRA 55 to reveal the transformation of the precipitations during heat treatment. All the specimens are cut parallel to the rolling direction.

Electrical resistivity measurements were performed using a conventional four point DC method. The measuring electrical circuit was powered from a current supply DH1720A-6, while the potential drops across the specimen and a standard resistance were directly read by RBH8321 data acquisition card on a personal computer. The specimens are cut parallel to the rolling direction. Specimens for tests are about 1 mm in thickness, and 50 mm length at least.

The microhardness was measured using a LEICA VMHT30M Vickers hardness tester at a load of 100 gf. The specimens for microhardness measurement were annealed at temperatures between 423~773 K for 1 h and water quenched.The tensile tests were conducted on a MTS 810 Materials Testing System. The specimens for the tensile measurement were annealed at different temperatures between 523~773 K for 1 h and water quenched.

Results and Discussion

To obtain the precipitation and coarsening processes of Al_3Sc phase, the alloy was pre-annealed at 873k in order that the Al_3Sc particles formed during the casting and rolling would solute into α-Al. More dispersive precipitates can be seen in the matrix of as-rolled Al-6Mg-0.2Sc alloy (Figure 1a) compared to the solution annealed alloy (Figure 1b). The precipitates are spherical or near-spherical (some of them were distorted during rolling). EPMA results indicate that they are Al_3Sc. The amount of Al_3Sc particles was decreasing with the aging time. Most of the Al_3Sc particles below 10 nm have soluted into the matrix.

There are in Figure 1 ellipse or quadrate phases precipitation besides Al_3Sc. The ellipse phase is $MnAl_6$ and quadrate phase is $(Mn, Fe)Al_6$ as determinate by EPMA and SEM, which agree with the results in literatures [17,18]. It was mentioned [19] that Fe can solute into $MnAl_6$, forming $(Mn, Fe)Al_6$, which could decrease the deleterious effect of Fe.

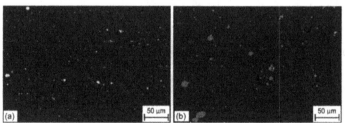

Figure 1. SEM micrographs of Al-6Mg-0.2Sc alloy:
(a) as-rolled; (b) solution annealing at 873 K for 120h

Figure 2 shows the SEM micrographs of the Al-6Mg-0.2Sc alloys aged at 573-723 K after solution annealing. One can see the variation of the dispersive precipitations during the aging treatment. The rate of Al_3Sc particles precipitation is faster at higher aging temperature, so to obtain approximate equal volume of Al₃Sc precipitates will need shorter time at higher temperature (Figure 2a, b, c). When aging at the same temperature, the quantity and the size of the Al₃Sc particles is increasing with the aging time (Figure 2d, e, f). The morphology of the Al₃Sc particles gets irregular shapes from sphere throughout the course of precipitation and coarsening. They commonly nucleate and precipitate on the grain boundaries, subboundaries and the dislocations, and finally precipitate in clusters at these places. The morphology of the Al₃Sc particles after aging is shown in Figure 3(a, b).

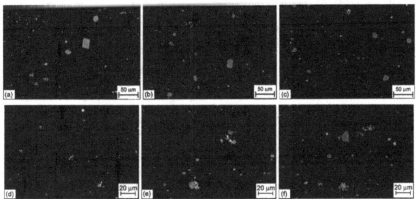

Figure 2. Microstructure of the solution annealed Al-6Mg-0.2Sc alloy during the subsequently aging: (a) 573 K for 24h; (b) 673K for 5h; (c) 723K for 2h; and aging at 623 K for (d) 12h, (e) 24h, (f) 144h

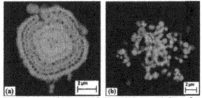

Figure 3. The morphology of the Al₃Sc particles after aging

Figure 4(a) shows the variation of the electrical resistivity of the as-rolled Al-6Mg-0.2Sc alloy in isothermal annealing in dependence on the treatment time. The electrical resistivity is increasing during solution annealing, while decreasing during the subsequently aging. The electrical resistivity of alloys can be expressed by Mathiessen Conductance Theory:

$$\rho = \rho_0 + \Delta\rho_{solution} + \Delta\rho_{precipitation} + \Delta\rho_{vacancy} + \Delta\rho_{dislocation} + \Delta\rho_{grain\ boundary} \qquad (1)$$

$\Delta\rho_{solution}$ has the most significant contribution to the resitivity change among these factors, and the contributions from other factors are decreasing as a sequence of $\Delta\rho_{precipitation}$, $\Delta\rho_{grain\ boundary}$, $\Delta\rho_{vacancy}$, and $\Delta\rho_{dislocation}$. During the solution annealing, $\Delta\rho_{solution}$ and $\Delta\rho_{vacancy}$ is increasing, so the value of the electrical resistivity becoming higher; During the subsequently aging, the Al₃Sc phase decomposes from the α-Al supersaturated solid solution, so $\Delta\rho_{solution}$ is decreasing and

$\Delta\rho_{precipitation}$ is increasing, and the value of $\Delta\rho_{solution}$ decreased is more than $\Delta\rho_{solution}$ increased, so the electrical resistivity is becoming smaller with the Al_3Sc precipitates formation [1].

Vickers microhardness is plotted against the annealing temperature in Figure 4(b). Both alloys were annealed at different temperatures for 1h. It is seen that the microhardness of Al-6Mg-0.2Sc plates are remarkably higher than those of Al-6Mg alloy. The Al-6Mg–0.2Sc alloy does not exhibit drastic decrease in the microhardness when the annealing temperature rise up to 723K, but the microhardness of the Al-6Mg alloy shows a significant decrease at 423K. This indicates that Al-6Mg alloy has already recrystallized before annealing, but recrystallization of Al-6Mg-0.2Sc alloy has been limited. The recrystallization temperature of Al-6Mg-0.2Sc alloy should be between 723~773K, which is much higher than Al-6Mg alloy.

Tensile strength of the studied alloys were tested after annealing at different temperatures for 1h (Table I). Both alloys exhibit decreases in strength and increases in elongations with elevating the annealing temperature. The ultimate tensile strength and the yield strength of Al-6Mg-0.2Sc alloy are much higher than those of Al-6Mg. The tensile strength of the as-rolled Al-6Mg-0.2Sc alloys increased by 80 and 160 MPa compared with Al-6Mg alloys. These differences become 55 and 120MPa after annealing at 523 K for 1h, 30 and 44 MPa at 773K. However, the least elongation of Al-6 Mg-0.2Sc alloy is also 8.5%, and the elongations of the alloys after annealing are over 15%. The reason is that fine dispersed Al_3Sc particles can tightly pin up the grain boundaries and the dislocations, so that the strength for the alloy is improved. Micrographs of fracture surface of the alloys during annealing are shown in Figure 5. Elongated dimples are presented for not only as-rolled but also annealed alloys. There are more dimples in Al-6Mg alloy than in Al-6Mg-0.2Sc alloys, and the dimples amount is decreasing and the dimples size becomes larger with the annealing temperature rising. The plasticity of Al-6Mg-0.2Sc is lower than Al-6Mg, because there are Al_3Sc particles forming and coarsening in the dimples in Al-6Mg-0.2Sc alloy, which increase the possibility of crack forming [20].

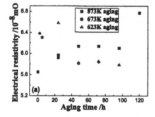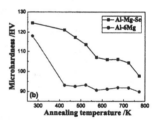

Figure 4. (a)The variation of electrical resistivity of the Al-6Mg-0.2Sc alloy during aging; (b) Microhardness of Al-6Mg and Al-6Mg-0.2Sc alloys with annealing, annealing time is 1h

Conclusions

1 During aging, there are Al_3Sc particles precipitating and coarsening in as-rolled Al-6Mg-0.2Sc alloys, the size and quantity of Al_3Sc particles are increasing, especially in grain boundaries; and the solution and precipitation of Al_3Sc particles influence the electrical resistivity of the alloy. The electrical resistivity is increasing when the Al_3Sc particles are soluted into the alloy, while it is decreasing during the precipitation aging.

2 By addition of 0.2 wt% Sc to Al–Mg alloy, the tensile strength of the as-rolled alloy increased by 80MPa and 160MPa , but the ductility still remains at higher level. After annealing, the tensile strength of the Al-6Mg-0.2Sc is much higher than Al-6Mg alloy.

3 The microhardness of Al-6Mg-0.2Sc plates are remarkably higher than those of Al-6Mg alloys not only in the as-rolled alloy but also after annealing. The recrystallization temperature of Al-6Mg-0.2Sc alloy is much higher than Al-6Mg alloy, The addition of 0.2 wt% Sc has effect on inhibiting the recrystallization of alloy.

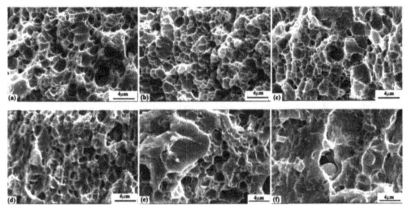

Figure 5. SEM micrographs of fracture suface during annealing: Al-6Mg: (a) as rolled, (b) 623K annealed, (c) 773K annealed; Al-6Mg-0.2Sc: (d) as rolled, (e) 623K annealed, (f) 773K annealed

Table I Tensile Properties of alloys after annealing

Alloy	Annealing temperature(K)	σ_b (MPa)	$\sigma_{0.2}$ (MPa)	δ (%)
Al-6Mg	as rolled	360	175	26.5
	523	355	160	26.5
	623	355	162	29
	673	360	162	31
	723	355	158	27.5
	773	350	156	29
Al-6Mg-0.2Sc	as rolled	440	335	8.5
	523	410	280	15
	623	395	230	18.5
	673	385	225	19.5
	723	385	210	22
	773	380	200	18

Acknowledgement

This work was supported by the National Natural Science Foundation of China (50501002 and 50771018), and the New Star Program for Science and Technology of Beijing City (2005B19).

145

References

1. Z.M. Yin et al., *Scandium and its alloys* (Changsha: Central south university press, 2007), 363-389.
2. T.N. Jin et al., "Effects of cooling rate on solidification behavior of dilute Al-Sc and Al-Sc-Zr solid solution," *Trans. Nonferrous Met. Soc. China*, 14(1) (2004), 58-62.
3. Jostein Røyset, Nils Ryum, "Kinetics and mechanisms of precipitation in an Al–0.2 wt.% Sc alloy," *Materials Science and Engineering A*, 396(2005), 409-422.
4. E.A. Marquis et al., "Composition evolution of nanoscale Al3Sc precipitates in an Al–Mg–Sc alloy: Experiments and computations," *Acta Material*, 54(2006), 119–130.
5. G.X. Xu et al., "Corrosion behavior of an Al-6Mg-Sc-Zr alloy," *RARE METALS* , 24(3) (2005), 246-251.
6. Y.A. Filatov, V.I. Yelagin, and V.V. Zakharov, "New Al–Mg–Sc alloys," *Materials Science and Engineering A*, 280(2000), 97–101.
7. O. Sitdikov et al., "Microstructure behavior of Al–Mg–Sc alloy processed by ECAP at elevated temperature," *Acta Materialia*, 56(2008), 821–834.
8. C. Watanabe , R. Monzen, and K. Tazaki, "Effects of Al3Sc particle size and precipitate-free zones on fatigue behavior and dislocation structure of an aged Al-Mg-Sc alloy," *International Journal of Fatigue*, 30(2008), 635-641.
9. F. Fazeli, W.J. Poole, and C.W. Sinclair, "Modeling the effect of Al3Sc precipitates on the yield stress and work hardening of an Al-Mg-Sc alloy," *Acta Material*, 56(2008), 1909-1918.
10. E.A. Marquis, D.N. Seidman, and D.C. Dunand, "Effect of Mg addition on the creep and yield behavior of an Al-Sc alloy," *Acta Material*, 51(2003), 4751-4760.
11. B.A. Parker, Z.F. Zhou, P. Nolle, "The effect of small additions of scandium on the properties of aluminium alloys," *Journal of Material Science*, 30(1995), 452-458.
12. N. Blake, M.A. Hopkins, "Constitution and age hardening of Al-Sc alloys," *Journal of Material Science*, 20(1985), 2861-2867.
13. C.B. Fuller, D.N. Seidman, D.C Dunand, "Mechanical properties of Al(Sc, Zr) alloys at ambient and elevated temperatures," *Acta Material*, 51(2003), 4803-4814.
14. D.N. Seidman, E.A. Marquis, D.C. Dunand, "Precipitation strengthening at ambient and elevated temperatures of heat-treatable Al(Sc) alloys," *Acta Material*, 50(2002), 4021-4035.
15. M.S. Kaiser et al., "Effect of scandium on the microstructure and ageing behavior of cast Al–6Mg alloy," *Mater Charact*, (2008).
16. M.H. Chen et al., "Superplasticity and superplastic forming ability of Al-Mg-Sc alloy," *Trans. Nonferrous Met. Soc. China* , 16(2006), s1141-s1414.
17. O. Sitdikov et al., "Grain refinement in a commercial Al–Mg–Sc alloy under hot ECAP conditions," *Materials Science and Engineering A*, (444)2006, 18-30.
18. Changrong Zhou, Qinglin Pan, Xinyu Liu, "The effect of minor Sc and Mn addition on microstructure and tensile property of Al-Mg alloys," *Material Herald*, 20(11) (2006), 147-149.
19. Shuisheng Xie, Jingan Liu, and Guojie Huang, *500 questions on the manufacturing technique of aluminum processing* (Beijing: Chemical industry publishing company, 2006), 32.
20. Z.J. Li, "Effect of complex alloying Sc, Zr, Ti on the microstructure and mechanical properties of Al-5Mg alloys," *Chinese rare earths*, 27(2006), 28-34.

146

ALLOY PREPARATION IMPROVEMENTS AT ALUMAR

Fernanda Silva[1], Jarbas Feitosa[1], Affonso Bizon[1], Sebastião Silva[1], Cristino Campos[1]

1Alumar Consortium; BR 135 km 18 Pedrinhas; São Luís, Maranhão,
P.O. Box 661 Zip Code 65,095-604, Brazil

Keywords: Aluminum Alloys, Furnace Preparation, Alloy Materials, Process Improvement

Abstract

To support Alumar's strategy of increasing aluminum alloys production, the process of preparing Al-Si alloy was investigated aiming at a higher efficiency on achieving the appropriate chemical composition, (measured as percentage of charges where the right chemical composition was obtained in the first sample), lower master alloys consumption and reduction on furnace turnaround time. The substitution of AlSr10% for AlSr15% on preparation of Al-Si modified alloys and tests regarding the use of a cage for master alloy addition are discussed. In addition, silicon pre-heating was evaluated aiming on a molten aluminum temperature loss reduction and consequently decrease on average furnace turnaround. The usage of copper scrap from anodes on Al-Cu alloys was also investigated showing considerable gains. This paper discusses the issues associated with "First Sample on Grade", and "Furnace Turnaround", for AlSi Alloy preparation, and examines the benefits that may be realized through a different approach in the alloying, pre-heating process and standard preparation practice change.

Introduction

The current demand on the aluminum market requiring a greater variety of Aluminum Silicon Alloys in order to attend the productive sector needs, challenged Alumar Ingot Plant to eliminate the gaps in the standard alloy preparation practices and process management system.

The identification of the major gaps was supported by the ABS (Alcoa Business System) - based on three principles: Make to Use, Eliminate Waste and People Make Happen – through the following tools:

- DMS (Daily Management System) for standardized furnace preparation practice
- IC&C (In control and capable), for process management

The optimization of Furnace Turnaround and Sample First Time on Grade were possible with a review of the operating practices, alloy material performance improvement and innovative process changes.

The main actions to achieve these objectives were:

- Alloy preparation DMS Board
- Furnace cleaning system
- Enrichment of Copper and Strontium alloying material
- Silicon preheating system
- The use of a cage for Strontium and Magnesium addition

The implementation and consolidation of these new practices were extremely successful resulting in the achievement of a 30% increase in alloy production between 2006 and 2008.

Process Management

Alumar's Ingot Plant is responsible for processing more than 1,230 tons of aluminum daily. Since 2005, Alumar's strategy involves the increase of higher value added production, specifically Aluminum alloys, without changing the industrial asset currently available.

In this way, the use of ABS techniques on the alloy production steps, mainly furnace preparation was crucial. The IC&C methodology intends to achieve process excellence by empowering employees to take responsibility for their work activities, with the objective to attend or even exceed customers' expectations, business requirements, through a capable production process.

The first steps to be taken are related to the identification of critical process and the main critical characteristics. Alloy preparation process was identified as critical and furnace turnaround and sample first time on grade as critical variables.

Studying the current situation of these two critical variables made it possible to bring up gaps and opportunities for improvement. Standardization of operational practices along with new technologies aiming the furnace preparation time and material loss reduction were the biggest contributors to optimize the preparation process.

DMS concept aligned with PDCA (Plan-Do-Check-Act) methodology was one of the facilitators for process standardization in all operating groups, supporting process improvements and production growth.

The Alloy Preparation Board (DMS) allows the operator to plan, execute and evaluate his own activities in the working area. All required information is available and in a dynamic way the operator interacts by searching and updating the DMS board.

The IC&C team works were also available at Alloy Preparation DMS Board, with the control charts for critical variables and a follow up sheet for tests and new implementations disclosure.

This DMS allows the operators to evaluate the results achieved, the performance by shift and the next steps for production and productivity increase, follow up and make them feel part of all the process improvements.

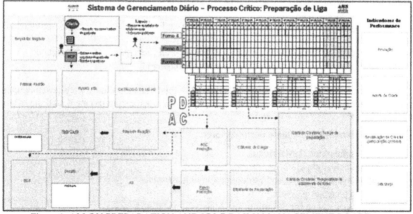

Figure 1. ALLOY PREPARATION AND IC&C DAILY MANAGEMENT BOARD.

Aluminum Alloying

Enrichment of Alloy Materials

The growing demand for better quality alloys along with productivity increase and saving has forced the searching of new technologies and materials. Evaluation of the current materials used in the alloy preparation regarding efficiency and reactivity aligned with reduction cost, made it possible to reach these improvements.

The main focus of this work is related to the use of Cu 99.8%, electrolytic copper, to replace AlCu75%, and AlSr15% as modified agent to replace AlSr10%.

Tests were carried out in a 45 ton furnace, taking into consideration temperature, stirring action, reaction time and proper time for addition.

The main gains obtained through the use of these new materials were:

- First sample on grade and furnace turnaround, by eliminating alloy batch correction, contributing to optimize both critical characteristics of the critical process.

- Material efficiency increase combined both, standard practices and addition techniques, by the use of a cage to alloy reactive elements such as Sr and Mg, submerging these elements while alloying, specially the AlSr master alloy, whose Sr content decreases when preparation time increases. Recovery obtained was between 80% and 85% for Sr, and 98% for copper rod scrap.

- Copper rod scrap being recycled internally eliminating a plant passive, with a cost reduction in storage, purchasing and transportation.

- Operators have to add less Sr waffle and copper bars for each alloy batch, less material handling and operator exposure which means a huge gain in ergonomics.

All these factors in conjunction with cost reduction regarding the prices and purity of these new materials end up in a saving of approximately US$ 300,000.

Alloy Materials Addition

The main objective of this work was to evaluate the impact of Sr and Mg addition practices. The use of a cage made possible the minimization of losses in Aluminum-Silicon alloy preparation.

In this study samples were taken during furnace preparation process in three distinct batches, with different temperature situation. The results were analyzed considering the Strontium recovery, preparation time and first sample on grade.

Figure 2. STEEL CAGE with ingots.   Figure 3. ADDITION WITH CAGE.

The main advantages of using the cage are:

- Total dissolution is obtained at the end of the addition process without any material loss due to oxidation and furnace atmosphere exposure.

- Maximization of surface area for contact between molten metal and master alloys.

The minimization of low temperature impact regarding furnace turnaround time and master alloy recovery was the main result of the cage use. In figure 4, it can be seen the Sr recovery in three different levels, all of them using cage for strontium addition but varying the batch temperature.

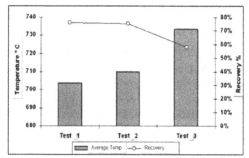

Figure 4. ALLOY MATERIAL RECOVERY with cage addition.

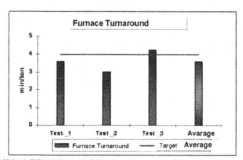

Figure 5. FURNACE TURNAROUND using cage for Alloy materials addition.

Furnace Cleaning Practices

Alumar's Casthouse receives more than 1200 tons of aluminum per day, with an average content of 15 kg/ton of bath tapped with metal, resulting in a daily input of 18 ton of bath into the Casthouse furnaces.

To minimize the impact of all this bath and the dross generated on alloy preparation, it was developed a furnace cleaning tool to be used in the fork lift, as shown in the figure 6. The main purpose of this tool is to facilitate the cleaning process, reducing time for cleaning and operator exposure to furnace temperature.

Figure 6. FURNACE CLEANING TOOL

A cleaning schedule was implemented with a distribution of the 6 furnaces for three operational groups; each one being responsible for the cleanliness of 2 furnaces. An audit is done every day to guarantee the cleaning efficiency and attribute to the responsible group a value for performance evaluation. This procedure was fundamental to support production growth without any investments on new furnaces and furnace lining, which led to a saving of US$216,000 per year.

Silicon Pre-heating

To guarantee the alloy production growth, efforts were concentrated to minimize the impact of temperature on furnace turnaround, especially on 11% Silicon Aluminum alloys. Studies were carried out to evaluate the benefits of silicon pre-heating before alloying.

A preliminary test was done in two different furnace charges. The silicon mass to be added was preheated on a 3 ton silicon furnace. After the heating period, silicon was added to the holding furnaces and furnace turnaround was measured during alloy preparation.

The test structure is shown on figure 7. The temperature of silicon after heating was close to 250°C, with an alloy casting temperature of 730°C. These two tests resulted in an average furnace turnaround reduction of 35 minutes per 45 ton of alloy prepared.

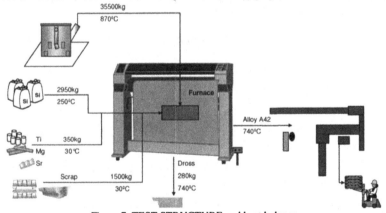
Figure 7. TEST STRUCTURE and heat balance

151

With the success of these two preliminary tests made it possible to extend the test period for more 20 days. The same structure and process settings were maintained and a statistical study was conducted to estimate the time reduction and time distribution on alloy preparation. The result from this study is shown on figure 8.

According to figure 8, it is possible to conclude that there is significant difference between alloy preparation with and without silicon preheating. There is a reduction on variability and an average decrease of 0.6min/ton prepared, which means more than 20 minutes reduction on a furnace preparation of 45 ton capacity. Other observations regarding temperature gains in electrical furnaces were noticeable, including the alloying potential of some master alloys, as titanium, which classified this practice as critical for ensuring the alloy production increase at Alumar.

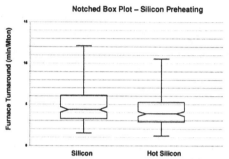

Figure 8. FURNACE TURNAROUND COMPARISON

Results

The initiatives on process management and alloying practices were fundamental for Alumar growth on alloy production. Figure 9 shows the improvement on furnace turnaround obtained as a result of all discussed changes and the stability obtained since September 2007. In the same line, figure 10 shows the improvements achieved on first sample on grade, joining alloy materials development, cage usage and process management.

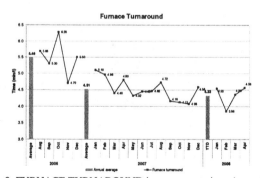

Figure 9. FURNACE TURNAROUND improvement since August, 2006

152

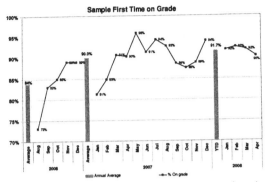

Figure 10. SAMPLE FIRST TIME ON GRADE improvement since August, 2006

Those results supported the production increase, shown at figure 11, without any significant implementation cost. In this scenario, Alumar Ingot Plant grew in the international market as a supplier of more than 100 different aluminum alloys. The operational expertise was successfully acknowledged with a score above 75% on the Ingot Plant Global Comparison and a cost saving close to US$3,5MM in 2007.

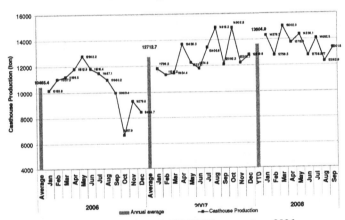

Figure 11. ALLOY PRODUCTION increase since 2006

Aluminum Alloys:
Fabrication, Characterization and Applications II

Modeling and Corrosion

Session Chair

Yansheng Liu

Analysis of Pulsed Forming Processes

Sergey F. Golovashchenko[1] and Nicholas Bessonov[2]

[1]Ford Research Laboratory,
Manufacturing & Proces Department, SRL
2101 Village Rd., P.O.Box 2053, MD3135,
Dearborn, MI 48121-2053

[2]Institute of the Problems of Mechanical Engineering, Russian Academy
of Sciences, 61 Bolshoy prospect V.O., St. Petersburg, 199178, Russia

Keywords: Pulsed, forming, aluminum sheet

Abstract

During recent decade, significant efforts were dedicated to increasing the amount of Aluminum Alloys and Advanced High Strength Steels in automotive parts in order to reduce the net weight of cars. Processes of pulsed forming, such as electromagnetic and electrohydraulic forming, are known to expand the capabilities of traditional stamping operations. Pulsed electromagnetic forming is based on high-voltage discharge of capacitors through a coil. An intense transient magnetic field is generated in the coil and through interaction with the metal work-piece, pressure in the form of a magnetic pulse is built up to do the work. An important requirement for both sheet metal material to be formed and for the coil material is good electrical conductivity. That often limits potential automotive applications of electromagnetic forming to parts formed out of Aluminum Alloys.

Pulsed electrohydraulic forming is a similar electrodynamic process, based upon high-voltage discharge of capacitors between two electrodes positioned in a fluid-filled chamber. There is a lot of similarity between electromagnetic and electrohydraulic forming in terms of short duration of the process, high velocities of the blank leading to its high strain rate deformation, similar power supplies, and the necessity to operate with high voltage in stamping environment. Significant advantage of electrohydraulic forming compared to electromagnetic forming is substantially broader choice of materials including variety of advanced high strength steels and aluminum alloys.

Specific attention will be dedicated to numerical modeling of pulsed forming processes. Suggested formulation included a simplified approach for calculating the pressure pulse and spending a major effort on modeling of contact interaction of the blank with the deformable die. Parameters of the pressure pulse can be refined at a later stage using already developed numerical models. Mild contact model based on introduction of acting-in-vicinity forces repelling the surfaces to be in contact is involved in calculating contact stresses. Dynamic deformation of both the blank and the die is done in 2D elastic-plastic formulation. The explicit integration procedure is preferred due to short duration of the process. The forming process is analyzed as a single-pulse operation, when the die is filled with the blank material as a result of one discharge of capacitors, and as a multi-pulse forming operation, where several discharges are involved. The results of numerical modeling indicated that the level of contact stresses can be lowered if several sequential discharges are produced. □

Introduction

In accord with the general trend of decreasing vehicle weight, light weight materials are being used more and more in automotive body construction. Improvement in formability would further promote the application of Aluminum Alloys (AA) and Advanced High Strength Steels (AHSS) for the production of body panels of automotive vehicles. Currently there are several approaches to achieving this improvement by employing alternative forming processes such as superplastic forming technology [1] or warm forming [2]. These technologies offer substantial improvement in formability of specific alloys, but also have the following issues: heating metal and stamping tools adds to the cost of the stamping process; thin blanks quickly cool down if they are not formed in isothermal conditions; and specific lubricants which tolerate high temperatures are required. Application of these advanced technologies is limited mostly by a family of non-aging AA. Other materials, like AHSS and aging AA, such as dual phase steels or 6xxx AA can not be formed by these methods.

In this paper the enhancement of formability by employing high-rate forming processes without preliminary heating of blanks will be discussed. Technologies of electromagnetic forming (EMF), electrohydraulic forming (EHF) and explosive forming were widely used in airspace applications and even in automotive applications [3,4], for example for electromagnetic joining of aluminum drive shafts.

The idea of using a pulsed electromagnetic field to generate forces sufficient to form metals and join tubular parts has been known since 1924 [5]. Pulsed electromagnetic forming is based on high-voltage discharge of capacitors through a coil. An intense transient magnetic field is generated in the coil and through interaction with the metal work-piece, pressure in the form of a magnetic pulse is built up to do the work. Obviously, high conductivity of the blank and coil material is required to have high efficiency of the process. In many cases, this requirement is in contradiction with the trend of using high strength alloys and steels. The most appropriate application of EMF technology in sheet metal forming is considered to be for AA, where substantial enhancement of formability can be achieved. The schematic of pulsed electromagnetic forming of sheet is shown in Fig.1.

After the blank is installed in the tooling, the capacitors are being charged through a high voltage transformer. As soon as required charging voltage is achieved, the control system sends a signal to the switch to close and the discharge of capacitors starts. It is desirable that the inductance of the coil would be substantially larger than the inductance of the equipment and connecting cables.

Fig.1. Schematic of electromagnetic forming process

The equipment and general features of EHF systems have been described by a number of authors [6-8]. In this process, a significant amount of electrical energy, (typically between 5 and 50kJ) is stored in a bank of capacitors which are discharged across a gap in water over a very short period of time. Stored energy W_0 is usually defined by capacitance C of the battery and charging voltage U_0 as

$$W_0 = CU_0^2/2. \tag{1}$$

Vaporization of a small volume of the liquid sets up a high intensity, high velocity shock wave in the liquid which can be directed to form sheet metal into a die. Since the liquid transmits the force, only a female die is required. The schematic of EHF is shown in Fig.2. The blank is initially clamped between the die surface and the EHF chamber. Electrodes are electrically insulated from the chamber in order to direct the discharge into the clearance between the electrodes. The die should be clamped to the chamber through the blank with substantial force, sufficient to withstand the pressure pulse delivered to the blank surface. EHF process is capable of forming variety of materials, including AA and AHSS. The critical issue is to provide sufficient strength and durability of the die in dynamic loading.

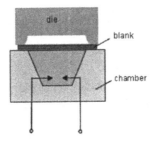

Fig.2. Schematic of electrohydraulic forming process

It is well known from previous publications that some benefits in pulsed forming technologies are from the impact of the blank into the die surface, such as formability improvement [9] and springback reduction [10]. One of the examples is improvement of formability in corner filling processes. The objective of current study is to understand the level of stresses generated in the die for a corner filling process, completed with one pressure pulse and with multiple pressure pulses.

Analytical Approach

Theoretical approach to modeling of pulsed loads applied to the blanks in EMF and EHF was previously discussed in [11, 8]. For EMF, it requires numerical solution of Maxwell equations in 3D formulation, while for EHF it is necessary to solve hydrodynamic problem to understand how pressure is transferred from the discharge channel to the blank. In this paper, no new progress is being reported in pulsed pressure calculation technique. Therefore, pulsed pressure is approximated by a sinusoidal function in the following form:

$$p = p_0\, e^{-\beta t} \sin(\omega t) \tag{2}$$

where ω=100 000, 1/sec β=10 000 1/sec, p_0 is an amplitude of the pressure pulse being varied to achieve the appropriate filling of the die with the blank material. The duration of the pressure pulse was limited to the time $t < \pi / (2\,\omega)$. For EMF it was dictated by the fact that the pressure is quickly diminished as soon as the blank moves away from the coil. For pulsed forming processes in general, it is rather typical that substantial portion of the deformation process the blank is driven by inertia, transforming the kinetic energy accumulated from the pressure pulse into the work of plastic deformation.

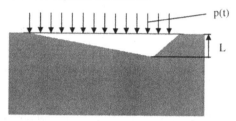

Fig.3. Schematic of the pulsed forming process

The simulation of the pulsed forming process was conducted using internally developed research code Solid 2D in two-dimensional formulation using a numerical solution of the solid mechanics equations of motion assuming the blank deformation to be plane strain.

$$\boxed{} \tag{3}$$

The blank material was assumed to be isotropic. The Prandtl-Reuss equations were employed.

$$\dot{e}_{ij} = \frac{\dot{S}_{ij}}{2G} + \lambda \; S_{ij} \; ; \; \dot{\sigma}_{o} = \frac{E}{1-2\mu} \; \dot{\varepsilon}_{o} \; , \tag{4}$$

The Von Mises yield criterion was used in the following form

$$S_{ij} S_{ji} = \frac{2}{3} \sigma_{y}^{\;2} \; , \tag{5}$$

where σ_y depends on plastic strain value. For practical calculations, the material workhardening was approximated as a piecewise linear function. The numerical integration procedure was based on explicit method.

The most popular method of modeling contact interaction is based on the geometrical analysis of mutual position of boundary nodes of each mesh [12]. At every integration step, it is being verified whether a boundary node of the blank has penetrated through the certain element of the surface mesh of the die. If it happens, it is necessary to make certain corrections bringing the node back on the surface of the die and let the node slide along die's surface instead of penetrating through this surface. A significant downside of this approach is in occasional penetration of the node through the surface. It usually happened due to the insufficient accuracy of the calculations. As soon as the blank's node penetrates into the die's surface, it is unable to return back and further calculations are useless. Therefore, a different approach was employed based on the idea of a mild contact. In mild contact formulation, the contact force is in inverse proportion to the distance between the interacting surfaces. From mathematical point of view, mild contact is some variable boundary of an unknown shape where we have to satisfy the non-penetration condition and equilibrium of forces between both surfaces. This approach is based on the introduction of acting-in-vicinity forces repelling the surfaces to be in contact. As a result, the surfaces do not come in contact, but stay at some very small distance from each other. The force is localized in a small neighborhood of mesh elements, and it increases to infinity when the distance between them is approaching zero. The absolute value of the force F is given by

$$F = \begin{cases} k\left(\dfrac{1}{h} - \dfrac{1}{h_0}\right) & \text{at} \quad h < h_0 \\ 0 & \text{at} \quad h \ge h_0 \end{cases}, \tag{6}$$

where h_0 is the width of the zone where the force is different from zero. The direction of the force F is aligned along h. Function (4) is to some extent arbitrary: it provides a repulsing force that increases to infinity as h tends to zero. The described algorithm allows taking into account the Coulomb friction and other contact effects. Indeed, when the interaction of the node with an object of the other mesh is computed, their mutual position and velocities are known. It is sufficient to find the friction directed along the tangent to the surface and opposite to the tangential part of the relative velocity vector.

The results of numerical simulation for a single-pulse and a multiple-pulse forming processes were compared for corner filling operation (Fig.3). The stress –strain curves for both the blank and the die was approximated as a linear function

$$\sigma_s = \sigma_y + A\,\varepsilon_{pl}\,, \tag{7}$$

where σ_y is material's yield stress; A is work hardening modulus; ε_{pl} is accumulated plastic deformation. For the die material the following properties were used: density $\rho = 7800$ kg/m3; shear modulus G=76.9 GPa, bulk modulus K=166 GPa, $\sigma_y = 620$MPa, A=1020 MPa. For the blank material: $\rho = 2735$ kg/m3; G=26.5 GPa, K=65.1 GPa, $\sigma_y = 195$ MPa, A=520MPa.

The vertical distance L from the targeted corner of the die to be filled to the blank surface (Fig.3) was considered to be a measurable quantitive parameter, characterizing to which level the blank has filled the die. The initial distance between the die corner and the flat blank (L before the beginning of the process) is 40mm. Another parameter characterizing the dynamic load on the die is the maximum contact stress applied to the die surface during the whole deformation process driven by one pulse of pressure.

Table 1. Characterization of the multi pulse forming process

	1st pulse	2nd pulse	3rd pulse	4th pulse	5th pulse
P_0	5 MPa	10 MPa	15 MPa	25 MPa	30 MPa
Max contact stress	10.9 MPa	24.1 MPa	45.0 MPa	79.2 MPa	88.7 MPa
L	32.3 mm	23.9 mm	15.05 mm	4.3 mm	0.9 mm

The multi-pulse process was modeled with five sequential pulses followed by the inertial movement of the blank and its contact interaction with the die. It was observed that the stiffness of the blank and the pulsed pressure necessary to deform it are growing while the blank is filling the die cavity. Therefore, the pressure amplitude is being increased from the previous to the next pulse. After the 5th pulse, the blank reached the die surface. The maximum contact pressure for the five-pulse process is 88.7 MPa. If a single-pulse process is considered, the necessary pulse amplitude in order to fill the die cavity is 40 MPa. In this case, the maximum calculated contact stress was 139 MPa. Obviously, for one-pulse process the contact load on the die is 57% larger than for a five-pulse forming process.

Conclusions

1. Developed numerical model provides the capability to define dynamic stresses in the die during high rate forming process.

2. Comparison of numerical results for one-pulse forming process and five-pulse forming process indicated that maximum contact stress is 57% higher for the one-pulse process.

References

1. J.C. Benedyk, "Superplastic Forming of Automotive Parts from Aluminum Sheet at reduced Cycle Times," *Light Metal Age*, 2002, no.6:28-31.

2. Betzalel Avitzur, *Handbook of Metal Forming Processes* (John Willey & Sons. New York, 1983) 1020.

3. P. Saha, "Electromagnetic forming of various aircraft components," *SAE paper 2005-01-3307*.

4. E.J.Bruno, *High velocity forming of metals*, (American Society of Tool and manufacturing Engineers, Dearborn, 1968), p.73-179.

5. P. Kapitsa, "A Method for Producing Strong Magnetic Fields," *Proc. Roy. Soc., Ser.A,* 105 (1924), 691-710.

6. V.N.Chachin et al., *Electro-hydro-pulsed treatment of metals* (Minsk: Nauka I Tehnika, 1987), 231.

7. V.S.Balanethiram, G.S. Daehn, "Hyperplasticity: Increased Forming Limits at High Workpiece Velocity," *Scripta Metallurgica et Materialia*, 30 (1994), 515-520.

8. S.Golovashchenko, V.Mamutov. "Electrohydraulic Forming of Automotive Panels," *Proceedings of 6th Global Innovations Symposium*, (TMS, 2005), p.65-70.

9. J.Imbert, M.Worswick, S.Winkler, S.Golovashchenko, V.Dmitriev: "Analysis of the increased formability of aluminum alloy sheet formed using electromagnetic forming," SAE Paper 2005-01-0082.

10. S.F.Golovashchenko, "Numerical and experimental results on pulsed tubes calibration," *Proceedings of a 1999 TMS Symposium "Sheet Metal Forming Technology"*, San-Diego,1999, 117-127.

11. S.Golovashchenko, N.Bessonov, R.Davies: "Pulsed electromagnetic forming of aluminum body panels," *Proceedings of 6th Global Innovations Symposium* (TMS, 2005) p.71-76.

12. S.F.Golovashchenko and N.M.Bessonov, "Numerical Simulation of High-Rate Stamping of Tubes and Sheets", *Paper presented at the International Seminar on Plates and Shells Theory*, Quebec,1996, Proceedings of the American Mathematical Society, 21 (1999) 199-207.

Aluminum Alloys: Fabrication, Characterization and Applications II
Edited by: Weimin Yin, Subodh K. Das, and Zhengdong Long
TMS (The Minerals, Metals & Materials Society), 2009

MODELLING THE HOMOGENIZATION HEAT TREATMENT
OF AA3003 ALLOY

Q. Du[1], W.J. Poole[1], M.A. Wells[2]

[1] Department of Materials Engineering
The University of British Columbia;
309-6350 Stores Road; Vancouver, BC, V6T 1Z4, Canada
[2] Department of Mechanical and Mechatronics Engineering
University of Waterloo;
200 University Avenue West; Waterloo, Ontario, N2L 3G1, Canada

Key words: AA3003, variational approach, nucleation, growth and coarsening

Abstract

The homogenization treatment of AA3003 involves the growth and dissolution of both inter-granular constituent particles and intra-granular dispersoids. It is a multi-scale problem involving both long-range diffusion (~10 μm) and short-range diffusion (~0.1 to 1 μm) of multiple alloying elements (Fe, Mn and Si). In this paper a comprehensive model is employed to simulate these phenomena. The model consists of a homogeneous nucleation model, 1D pseudo front tracking method for inter-granular constituent particles and a multi-precipitate growth model for dispersoids. The multi-precipitate model is developed based on a variational approach and it can capture the soft impingement and the influence of precipitate size distribution on the overall precipitate kinetics. The abilities of the multi-precipitate growth model are demonstrated by comparing it with existing models in the literature. The comparison of the simulation results with the experimental measurement for an industrial practice of this homogenization treatment is also conducted.

Introduction

AA3003 alloy is a Mn-containing commercially important aluminium alloy and has been used in many industrial sectors. Homogenization treatment prior to subsequent fabrication process is an important step in its whole manufacturing route. The evolution of microstructure during industrial homogenization treatment has been extensively studied from both experimental [1,2] and modeling points of view [3,4,5]. The starting point for the microstructure evolution path is the as-cast microstructure which mainly consists of equiaxed globular primary phase grains with about 3% inter-granular second phases as identified by Li and Arnberg [2]. The characterizing microstructure feature for the primary phases (i.e. the average secondary dendrite arm spacing) is ~ 20 μm. The secondary phase consists of a large quantity of rod like and plate like particles hereafter referred to as constituent particles. Most of them have been identified to be Al$_6$(Mn, Fe) with a size of ~ 1 μm and only a small fraction are determined to be α-Al(Mn,Fe)Si. Under industrial cooling conditions, severe microsegregation exists, i.e, the primary phase is super-saturated and the compositional profile across a secondary arm is non-uniform. The homogenization heat treatment causes the as-cast microstructure to equilibrate through two primary precipitation mechanisms. One of them is the growth and transformation of constituent particles, in which the diffusion at the scale of secondary dendrite arm spacing, hereafter referred as long-range diffusion, plays a very important role. This long range diffusion transports the alloying component (Fe, Mn and Si) through the supersaturated primary phase matrix to the interdendritic region, which first transforms $Al_6(Mn, Fe)$ to α-Al(Mn,Fe)Si and then grow the latter phase. The other precipitation mechanism is the nucleation, growth and coarsening of smaller α-Al(Mn,Fe)Si particles, hereafter referred to as dispersoids, in the intra-granular region which depletes the super-saturated matrix. According to the experimental results of Li and Arnberg [1], the size of the dispersoids is between 10 to 200 nm and number density ranges from 10-1000/μm^3. For these dispersoids, the diffusion at the scale of a fraction of micron, which is the inter-dispersoid spacing (hereafter referred as short-range diffusion) plays an important role during their growth and coarsening. Obviously the long- and short-range diffusions are not separate matters, and instead, tightly coupled. The growth/coarsening of inter-granular constituent particles requires the long-range transport of Fe and Mn, which are also in demand for the short-range growth/coarsening of inter-granular dispersoids. The long range diffusion of Si is also important for the transformation from $Al_6(Mn, Fe)$ to α-Al(Mn,Fe)Si phase. As suggested by Li et al. [5], the transformation from $Al_6(Fe,Mn)$ to α-Al(Mn,Fe)Si is controlled by the available content of Si in the surrounding Al matrix. This transformation consumes Si in the alloy while the Si concentration in the solid solution has a strong influence on the equilibrium solid solution level of Mn and Fe, in turn, influences intra-granular dispersoids precipitate. The volume fraction evolution of α-Al(Mn,Fe)Si dispersoids tends to decrease while the fraction of α-Al(Mn,Fe)Si constituent particles increases during the coarsening stage because they are essentially the same precipitate but the constituent particles are larger and have a size advantage (i.e. the Gibbs-Thompson effect). Therefore, it is very important to have a multi-scale model which could take into account the interaction of long and short range diffusion of all alloying components.

Lok and Miroux attempted to solve this problem by developing an analytical model [4]. However their model is limited to a single diffusing component (Mn) and does not take into account the transformation kinetics from $Al_6(Mn, Fe)$ to alpha-Al(Fe,Mn)Si. Another comprehensive model was proposed by Gandin and Jacot [3]. Their approach could quantitatively model the interaction between long and short range diffusion. It takes into account multi-component diffusion and the transformation kinetics from $Al_6(Mn, Fe)$ to α-Al(Fe,Mn)Si. It is able to predict particle size distribution. However one of the key elements in this model, the growth model for a single precipitate, which is based on the assumption that the precipitates remain spherical and are surrounded by a steady-state diffusion field as given in Aaron et al. [6], requires examination. This growth equation is only applicable to the precipitate kinetics under lower supersaturation, i.e., vanishing precipitate volume fraction. Marqusee and Ross [7] concluded that for finite volume fractions, the effect of competition among particles causes a deviation whose first order goes as the square root of the volume fraction in the coarsening rate. Recently a technique based on a variational approach was proposed. It is a very efficient technique with modest accuracy by employing several physical-based parameters describing the compositional profile surrounding a precipitate. It has been demonstrated that it could capture the interaction of the diffusion fields of two growing precipitates. Documenting the model is still in process. The extension of this technique to multi-precipitate growth will be documented in a separate paper. The aim of this investigation is to apply this approach to simulate precipitation kinetics of AA3003.

Model Description

The model work consists of three parts: a nucleation model, a 1D Pseudo Front Tracking (PFT) model for the long range diffusion and a Multi-Precipitate Growth (MPG) model for the short range diffusion. The 1D PFT and MPG are linked with each other through a kind of average composition variable in a volume element. In addition, these three models have to be coupled with the CALPHAD software Thermo-Calc to be applied to commercial alloys. At the current stage, the coding of the 1D PFT and MPG as well as their coupling with Thermo-Calc is complete. The linking of the two models and the implementation of the nucleation model is still in process. Although the 1D PFT and MPG models are used at different scales, it should be noted that both of the models do the same thing: simulating diffusion-controlled phase transformation. The 1D PFT model is considered an accurate numerical solution while the MPG model is more approximate but with the ability of handling many precipitates.

Nucleation Model

A classical nucleation model is adopted. The nucleation rate is calculated by the following equation:

$$\frac{dN}{dt} = N_0 Z \beta^* \exp(-\frac{\Delta G^*}{kT}) \exp(-\frac{\tau}{t}) \tag{1}$$

Where Z is the Zeldovich factor with $Z = \frac{v_{at}^P}{2\pi R^*}\sqrt{\frac{\gamma}{k_B T}}$. β^* is the condensation rate of solute atoms in a cluster of critical size with $\beta^* = \frac{4\pi R^{*2} DX}{a^4}$. τ is the incubation time for nucleation with $\tau = \frac{4}{2\pi\beta^* Z^2}$. ΔG^* is the energy required to form a critical nucleus of radius R^* with $\Delta G^* = \frac{16\pi\gamma^3}{3\Delta G_v^2}$. $R_{k_B T}^*$ is the nucleus size. $R_{k_B T}^* = R^* + \frac{1}{\sqrt{\pi Z}}$

The physical properties required to compute nucleation rate are N_0, the number of nucleation site per unit volume, γ, the interfacial energy, D, the diffusivity of alloying component in matrix, ΔG_v, the driving force for precipitate per unit volume, X the matrix mean solute atom fraction, v_{at}^P is the mean atomic volume within precipitates and a the lattice parameter of matrix phase.

1D Pseudo Front Tracking Method

A 1D PFT model is adopted here to describe the long range diffusion. The details of this model can be found in [8,9] and only a brief introduction is given here. The model is based on the following assumptions: (i) the composition of the intergranular region is uniform (ii) the intergranular region is locally in thermodynamic equilibrium and all phases have uniform concentrations and (iii) the long range diffusion affects the intergranular region in a uniform manner. The solution of the problem is based on the finite volume method. A volume element or cell of the mesh has three possible states: matrix, intergranular mixture, or interface. A layer of interface cells always separates the matrix and the intergranular mixture. An explicit formulation using the finite volume method provides a variation of solute concentrations in each cell with matrix states.

Multi-Precipitate Growth Model

The multi-precipitate growth model adopted here is derived using the variational approach-based precipitate growth model described elsewhere. This model is able to solve the moving boundary diffusion problems efficiently while keeping reasonable computational accuracy. This model could be applied to simulate the growth and coarsening of up to ~1,000 precipitates within reasonable computation time. Therefore it is suitable to be used for predicting precipitate size distribution. Having taken into account soft impingement, it provides an alternative to the method proposed in [10,11]. The basic assumption in this model is that the compositional profile surrounding each precipitate is spherical symmetric and has the form:

$$\tilde{c}(r,t) = c_m(t) + \nabla c(t)(a(t)-r)(\frac{b(t)}{r}) + \Delta c(t)\left(\frac{a(t)-r}{a(t)-b(t)}\right)^2 (\frac{b(t)}{r}), b(t) \leq r \leq a(t) \quad (2)$$

It consists of three terms which are: constant far field composition term, a first order inverse linear term representing outwards constant flux and a quadratic equation term representing the localized compositional gradient in the front of the migrating interface respectively. The improvement of this compositional profile to the one used in steady state growth model in [3, 4, 10, 11] can be seen by comparing Eq. (2) with their the widely adopted first order inverse linear compositional profile which has the form:

$$\tilde{c}(r,t) = c_m(t) - (c_m(t) - c_{eq}(t))\frac{r}{x} \quad (3)$$

First, an additional degree of freedom, i.e., $a(t)$, the boundary layer size has been added to the first order inverse linear term. While in Eq. (3), the boundary layer size is implicitly assumed to be infinite no matter what the precipitate size is, which has prohibited a rigorous description of soft impingement. The cost of introducing the additional boundary layer size parameter is the need to prescribe boundary conditions for compositional boundary layer. For isolated growth, the following equations are applied

$$\nabla \dot{c}_i = 0, \dot{c}_{mi} = 0, J_i^0 = 0 \quad (4)$$

During interrupted growth, the following equations are applied.

$$\sum_i^N a_i^{Coordinate\_Type} = 0, \sum_i^N a_i^{Coordinate\_Type-1} J_i^0 = 0, \dot{c}_{m,i} = \dot{c}_{m,i+1}, J_i^0 = D\left(\frac{\partial c}{\partial r}\right)_{r=a_i} \quad (5)$$

In which *Coordinate_Type* is 3 for spherical coordinate and 1 for Cartesian coordinate.
Second, a quadratic equation term in the compositional profile had been added reflecting that the composition gradient in the front of a moving interface is localized while in Eq. (2) the composition profile is smeared and has a uniform outward radius flux, which does not allow a proper description of growth kinetics.
The evolution of the two DOF, $c_m(t), c_{eq}(t)$ in Eq. (3) could be determined easily from a local equilibrium assumption and mass balance, while the evolutions of all of DOFs present in Eq. (2) could be derived by employing the variational principles. The variational principal in its most simple form can be written as:

$$\Pi = \Psi + \dot{G} \quad (6)$$

where Ψ is the dissipation potential which represents the rate of work done by the kinetic processes, and $\dot{G}$ is the rate of change of Gibbs free energy which is the thermodynamic driving force for the evolution. The dissipation potential for diffusion of a single component in a system of volume *V* is given by:

$$\Psi = \int_V \frac{j_m j_m}{2\mathcal{D}_V} dV \quad (7)$$

where j_m is the volumetric flux, $\mathcal{D}_V$ is the bulk diffusivity. j_m is a function of compositional profile, i.e, Eq. 2, as stated by Fick's first law. Therefore Ψ, the dissipation potential, is a solely determined function of all of DOFs present in Eq. 2. The Gibbs free energy has volumetric and surface contributions, i.e.:

$$G = \int_V g dV + \int_A \gamma dA = \int_V g dV + \sum_i^N 4\pi b_i^2 \gamma \quad (8)$$

166

where γ is the interfacial energy per unit area and A is the interfacial area between different phases, and g is the volumetric (phase) energy density. Since g is only dependent on composition besides temperature and pressure, its volumetric integration only depends on the DOFs as well. The variational principle states that the kinetic path is chosen in such a way that the above functional is rendered stationary, i.e.: $\quad \delta\Pi = 0$ $\qquad\qquad$ (9)
Therefore the governing equation for all the DOFs of the system could be derived from the above equation.

Results and Discussions

A single dispersoid growing in a 2 μm domain for an AA3003 (Al-Fe-Mn-Si) FCC matrix upon a typical homogenization heat treatment is simulated by both 1D PFT and Multi-Precipitate Model. The purpose of performing these two calculations is to verify the result of MPG by 1D PFT which is considered as an accurate numerical model. As shown in Fig. 1(a), the two models give almost identical dispersoid fraction evolution curves for the given temperature time history also shown in this figure. That is very encouraging considering the fact that the computation time of MPG is only a small fraction of the one of 1D PFT. Both of the two models predicts a rapid increase in dispersoid fraction from 4 hours of heat treatment, then a slight decrease after 5 hours, which is due to the higher solubility of Fe and Mn in the matrix phase (i.e., lower equilibrium α-Al(Mn,Fe)Si fraction) at 873.15 K. During the isothermal holding stage, the dispersoid fraction is constant. This constant value does not make sense in the context of industrial heat treatment of this alloy because long range diffusion would transport Fe, Mn and Si to the inter-granular region and grow the big constituent particles at the cost of small dispersoids, which is not considered in the two calculations. During the cooling stage, dispersoid fraction continues to increase due to the lower solubility of alloying components at lower temperature.

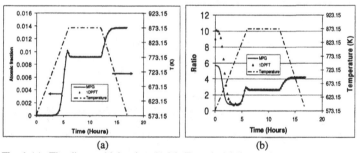

(a) $\qquad\qquad\qquad\qquad\qquad$ (b)

Fig. 1 (a) The dispersoid fraction (b) Mn/Fe ratio in dispersoid upon a typical homogenization heat treatment by 1D PFT and MPG for Al-Fe-Mn-Si alloy.

The parameters used in the calculations include thermodynamic functions (from Thermocalc), surface energy and diffusion coefficients, which is taken from or set as TTAl6 database, 0.127 J m^{-2}, and $D_{Fe}^{Fcc} = 0.363 \times \exp(-\frac{214000}{RT}), D_{Fe}^{Fcc} = 0.0135 \times \exp(-\frac{211500}{RT}), D_{Fe}^{Fcc} = 0.0000138 \times \exp(-\frac{117600}{RT})$. respectively.

It is also interesting to compare the predicted Mn/Fe ratio of the dispersoids with the measured one. EDS measurements by Li and Arnberg [1] indicate that the ratio is very high at low temperatures and it decreases as the temperature increases. The average Mn/Fe ratio of the dispersoids precipitated at 500 °C, 560 °C and 600 °C are 18.5, 3.8 and 2.5 respectively. At the same time, the Mn/Fe ratio of the primary particles has increased from 0.57 in the as-cast state to 0.59 after heating to 600 °C. This implies that there is some diffusion of alloying elements between the Mn-rich dispersoids and the Fe-rich primary particles, which is not taken into account in the two calculations. A similar trend is observed in the calculations during the heating stage as shown in Fig. 1(b). Again the calculation results during isothermal holding and cooling are not comparable with the measurement due to the neglect of long range diffusion in the calculations.

The growth and coarsening of two dispersoids with various sizes during a typical homogenization heat treatment are simulated using the MPG model. Fig. 2 shows the Mn compositional profile (in atomic fraction) surrounding the smaller dispersoids during isolated growth, soft impingement and coarsening. The x-axis starts at the precipitate/matrix interface and extends to the end of the boundary layer. It should be noted that the black curve in the Fig. 2 has a very small negative slope, which cannot be seen due to the scale used in the figure. This calculation confirms the models ability to simulate soft impingement, which is not dealt with in KWN model.

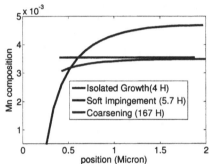

Fig. 2 The compositional profile of the smaller dispersoid at various stages.

Conclusions
A new comprehensive model is employed to simulate the homogenization heat treatment of an AA3003 alloy and preliminary results are presented in this paper. It is concluded that the model, subject to some improvement, may be a better alternative than the widely used mean field model.

References

1. Y.J. Li, L. Arnberg, Acta Materialia 51 (2003) 3415-3428.
2. Y.J. Li, L. Arnberg, Materials Science and Engineering A347 (2003) 130-135.
3. Gandin Ch.-A, Jacot A, Acta Materialia 55 (2007) 2539-2553.
4. Z.J. Lok, A. Miroux and S. van der Zwaag, Materials Science Forum Vols. 519-521 (2006) 443-448.
5. Y.J. Li et al, Materials Science Forum Vols. 519-521 (2006) pp 297-302.
6. H.B. Aaron, D. Fainstein and G.R. Kotler, Journal of Applied Physics, Vol. 41, No. 11, 1970, 4404-4410.
7. J. A. Marqusee and J. Ross, J. Chem. Phys. 80 (1), 1 January 1984. pp. 536-543.
8. A. Jacot, M. Rappaz, Acta Materialia 50 (2002) 1909-1926.
9. Q. Du, A. Jacot, Acta Materialia 53 (2005) 3479-3493.
10. O.R. Myhr, O. Grong, Acta Materialia 48 (2000) 1605-1615.
11. Robson JD, Acta Materialia 52 (2004) 4669-4676.

WELDING TECHNIQUES AND CORROSION BEHAVIOR OF 5XXX ALLOY FOR MARINE STRUCTURAL APPLICATION

Zhengdong Long, Subodh Das, [1]
J. Gilbert Kaufman, Shridas Ningileri, Yufu Wang [2]

[1] Center for Aluminum Technology, University of Kentucky, 1505 Bull Lea Rd., Lexington KY
[2] Secat, Inc., 1505 Bull Lea Road, Lexington KY

Key Words: Friction Stir Welding, Intergranular Corrosion, AA5456, Naval Ship

Abstract

Al-Mg 5xxx aluminum alloys are broadly used in naval ship structures due to their superior strength-to-weight ratio and generally excellent salt-water corrosion resistance. However, the strength and corrosion properties are significantly affected by joining with the traditional gas metal arc welding (GMAW) process. The large heat affected zone resulting from the GMAW has relatively low strength and is susceptible to intergranular corrosion. The Navy has experienced intergranular corrosion (IGC) corrosion of Al-Mg alloys used in long-term exposure in equatorial environments. The friction stir welding (FSW) technique provides less heat input, potentially providing improvement in this condition, and therefore was investigated in this study. The welding efficiency, microstructure, microhardness and corrosion resistance of FSW and GMAW joints in a representative 5456-H116 plate were compared. The FSW provides higher weld strength efficiency and improved corrosion resistance.

Introduction

In today's U.S. Navy ships, most of the superstructure above the main deck is made of aluminum, and many other aluminum components are found throughout ships. Since the use of aluminum in place of steel saves weight, an ever-increasing amount of aluminum use is expected, especially as new generations of high speed surface ships replace traditional vessels. [1] However, corrosion problems have limited the application of aluminum alloys with more than 3 wt% Mg for long-term service in naval superstructures. The Navy has experienced intergranular corrosion failures in superstructures after about twenty years of service in hot, humid environments.

It is recognized that the poor corrosion resistance of high-Mg-containing aluminum alloys is the result of precipitation of networked Beta phase (Al_3Mg_2) along grain boundaries after long term exposure at temperature higher than 65^0C. [2] The formation of beta phase precipitation along grain boundaries leads to "sensitization" to intergranular corrosion attack. The beta phase is more anodic than the matrix or depleted zone, and thus corrosion attack will occur along grain boundaries containing beta phase buildup.

Some of the corrosion problems have occurred in regions where 5xxx alloy plates have been joined using the traditional gas-metal-arc welding (GMAW) process. GMAW generates large heat-affected zones that are susceptible to the intergranular corrosion attack. In this paper, the use of friction stir welding (FSW), a thermomechanical process just below the melting point of the materials, [3, 4, 5, 6, 7] to join 5xxx alloy plates was investigated to determine if an improvement in weld microstructure would reduce intergranular corrosion of the weld and heat-affected zones, thereby improving the long-term mechanical efficiency of welded components.

Experimental Approach

For this study, FSW and GMAW joints were produced transversely in 6-mm thick commercially produced 5456-H116 plate. The chemical composition of this 5456 lot was Al - 5.1Mg - 0.75Mn - 0.13 Cr wt%. FSW was conducted on a Nova-Tech C-frame friction stir welding machine. Figure 1 schematically shows the setup of the welding and pin. The 125-mm long by 250-mm wide strips of plates were friction stir welded (FSW) in butt-joint configuration, with the weld running across the rolling direction. A 300-rpm spindle speed and 125-mm/min. travel speed were used. The diameters of the pin tip and end are 6.4 and 7.7 mm respectively with height of 6 mm. The pin has three threads and the diameter of shoulder is 18 mm.

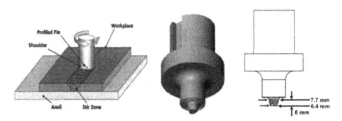

Figure 1: Schematics of Friction Stir Welding process

GMAW was made in the same configuration of the 5456-H116 plate for direct comparison with the FSW joints. The GMAW was shielded using argon gas at 35cfh gas flow rate. The filler wire was 1.14 mm in diameter 5556 (Al - 5.2Mg – 0.75Mn – 0.12Cr – 0.12Ti wt%) wire. The plate was welded in two passes with 41m/min travel speed.

Exfoliation and intergranular corrosion tests, tensile tests, microhardness tests, and microstructural examinations were made of the FSW and GMAW joints and of the parent metal for comparison. The intergranular corrosion tests were performed under the guidelines from ASTM G67, which is the standard test method for determining the susceptibility to intergranular corrosion of 5xxx series aluminum alloys by mass loss after exposure to nitric acid. The samples size was 50 mm long and 6 mm wide of the full plate thickness. The sample weights were measured before and after the immersion in the nitric acid for 24 hours at 30°C.

ASTM B557 is the standard tensile test method for aluminum alloys, compatible with ASTM Method E8 covering tensile tests of all metals. The tensile tests of the welds and parent metal were made with full-thickness 12.5 mm wide sheet-type specimens taken across the welds and, for the parent metal, in the transverse rolling direction. Two specimens were tested and the average value was used for analysis. All the welded samples were machined flat prior to tensile testing. Microhardness tests were made with a Vickers Microhardness Tester, Mitutoyo HM-122, in accordance with ASTM Standard Method E 384-07. The loading force is 4.9 Newton and dwell time is 15 seconds.

As a further indicator of the long-term corrosion resistance, samples of each weld and of the parent metal were given high-temperature exposures of 4 days of exposure at 117°C, estimated to be approximately equivalent to 25-30 years at 65°C based upon the Larson-Miller time-temperature parameter (LMP). [8, 9] The grain structure was revealed by using Barker's etchant for 60 seconds.

Results and Discussion

1. Tensile properties

Table 1 shows the cross-weld tensile properties of the FSW and GMAW joints compared with those of the parent 5456-H116 with and without thermal exposure in the transverse rolling (TD) direction. The FS weld has significantly higher ultimate strength (+16.2%) and yield strength (+10.7%) than the GMA weld. The FS weld also shows markedly less reduction in strength from the parent plate than the GMA weld. For the FS weld, the ultimate strengths are almost same before and after welding, compared to 13.5% reductions for the GMA weld. The reduction in yield strengths were 6.7%, compared to 15.7% for the GMA weld. These comparisons illustrate one direct benefit of the FSW for marine structures. It is recognized that these percentages may differ depending upon sheet or plate thickness and weld configuration. The mechanical property advantage of FSW technique provides potential weight reduction benefit for the civilian and naval ship structures.

Table 1: Tensile properties of 5456-H116 parent metal and FSW and GMAW joints

CONDITION	UTS,	TYS,	Elongation %				
As-fabricated, Specimen-1	345	199	7.4	Reduction - Compare			
As-fabricated, Specimen-2	342	202	7.8	to As-Fabricated			
As-fabricated, Average	344	201	8	UTS	TYS		
As-fabricated + aged, Specimen-1	334	174	10.8				
As-fabricated + aged, Specimen-2	328	180	11.0	-3.6%	-11.7%		
As-fabricated and aged, Average	331	177	10.9				
As-fabricated + GMAW, Specimen-1	301	172	--			Ratio of FSW to	
As-fabricated + GMAW, Specimen-2	294	166	--	-13.5%	-15.7%	GMAW	
As-fabricated and GMAW, Average	297	169	--			UTS	TYS
As-fabricated + FSW, Specimen -1	340	188	--				
As-fabricated + FSW, Specimen -2	350	186	--	0.4%	-6.7%	16.2%	10.7%
As-fabricated + FSW, Average	345	187	--				

2. Microstructure

As a solid status joining process, FSW usually has three zones, i.e. stir zone (also nugget), thermo-mechanically affected zone (TMAZ) and heat-affected zone (HAZ). The stir zone is a region of heavily deformed material that roughly corresponds to the location of the pin during welding. The TMAZ occurs on either side of the stir zone. In this region the strain and temperature are lower and the effect of welding on the microstructure is correspondingly smaller. HAZ is common to all welding processes.

Figure 2 shows a comparison of the macrostructures of the FSW and GMAW joints. The welding zone of GMAW and stir zone of FSW are clearly visible. The FSW stir zone in the center of the thickness is about 10 mm wide, which is wider than that in the bottom of joint and narrower than that in the top of joint. Advancing and retrieving sides can be clearly observed. The boundary between the FSW weld zone and base metal is much sharper for advancing side than that in retrieving side. The difference is caused by the semi-solid mass flow behavior during the welding. The relatively sharp boundary in advancing side occurs where the pin translational movement is the same as the rotational pin motion. However, TMAZ, HAZ zones of FSW and HAZ of GMAW can not be clearly identified by macrostructure.

Figure 3 gives the grain structures of FSW and GMAW joints in the locations of stir zone/welding zone, TMAZ and HAZ as well as parent metal. A clear boundary between welding zone and HAZ can be observed for GMAW. For FSW, the TMAZ with characterization of deformation and non-recrystallization can not be clearly identified. A less than 10 um small transaction zone between stir zone and HAZ could be the TMAZ. The grain size in the stir zone of FSW joint is much finer than that in welding zone of GMAW joint, finer even than the very fine structure in the parent metal. The grain structure is so fine in the FSW stir zone that it can not be clearly observed under normal optical microscopy.

The stir zone of FSW process experiences mostly mechanical flow of the base metal, mostly slightly below the melting temperature, so no obvious as-cast microstructure features can be observed. On the other hand, the GMAW welding zone experiences a typical re-melting and re-solidification process, leading to a weld zone with an as-cast microstructure. The larger size of the secondary precipitate in the GMAW heat affected zone is the result of the exposure to the relatively higher temperature generated by GMAW than that generated by FSW.

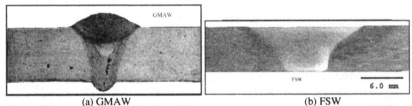

(a) GMAW   (b) FSW
Figure 2: Macrographs of the cross section of (a) GMAW and (b) FSW in 5456-H116 plate

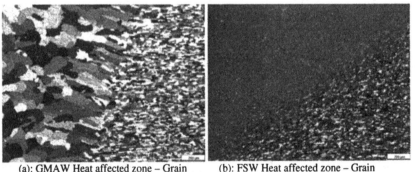

(a): GMAW Heat affected zone – Grain   (b): FSW Heat affected zone – Grain

(c): GMAW weld zone - Grain   (d): FSW stir zone - Grain
Figure 3: Microstructure of FSW and GMAW processed sample

3. Microhardness

Microhardness measurements made across the two joints are plotted in Figure 4. The typical "V" shaped microhardness plot across the welding zone can be observed for the GMAW processed plate. The minimum microhardness of 81 Hv was observed at the center of welding zone. The microhardness gradually increases from the minimum of 81 Hv at welding zone center to about 100 Hv at about 20 mm away from the center. As shown in Figure 2, the welding zone at plate center is about 4.5 mm. Therefore, the heat affected zone is about 18 mm for each side of the welding zone.

A "W" shaped microhardness plot is observed across the FSW in plate center. As consistent with Figure 2, the right side and left side of Figure 4 are corresponding to the advancing and retrieving sides respectively. A high microhardness of about 100 Hv, which is close to the microhardness of parent metal, can be noticed for about 9 mm wide in the center of welding zone. This high microhardness zone is the FSW stir zone based on the fact that the FSW stir zone is about 9 mm at plate center. There are two low microhardness zones, with lowest microhardness of 86 Hv, between stir zone and each side of parent metal. These two low microhardness zones are HAZ zone with 15 mm wide.

There are no large difference between advancing side and retrieving side. In comparison with GMAW, FSW joint's lowest microhardness of about 86 Hv is higher than GMAW joint's lowest microhardness of 82 Hv. The 15 mm FSW heat affected zone is less than 18 mm GMAW heat affected zone.

173

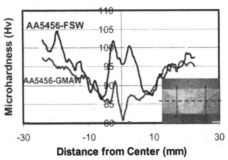

Figure 4: Microhardness comparison of FSW and GMAW joints in 5456-H116 plate

The stir zone of FSW process experiences mostly mechanical flow of the base metal, mostly slightly below the melting temperature, so no obvious as-cast microstructure features can be observed. On the other hand, the GMAW welding zone experiences a typical re-melting and re-solidification process, leading to a weld zone with an as-cast microstructure. The larger size of the secondary precipitate in the GMAW heat affected zone is the result of the exposure to the relatively higher temperature generated by GMAW than that generated by FSW.

Based on the above microstructure analysis, it seem that the high microhardness FSW welding zone reflects thermo-mechanically processed 5456 material properties, while for the GMAW joint, the lowest values are at the center of the weld zone reflecting the completely annealed cast 5556 filler alloy.

5. Intergranular corrosion resistance

Susceptibility to intergranular corrosion of aluminum alloys by mass loss after exposure to nitric acid was examined based on ASTM G67. Based on the ASTM G67, it is classified as sensitized material if the weight loss is more than 160 mg/cm^2. The material is not sensitized if the weight loss is less than 15 mg/cm^2. The intermediate sensitization range is defined as weight loss between 15 and 160 mg/cm^2.

Figure 5 gives the intergranular corrosion resistance testing results based upon mass loss in ASTM G67. The left three bars shows comparison of weight losses among FSW, GMAW, and parent metal. While the weight loss is higher after welding either by FSW or GMAW, the weight loss for FSW is smaller by about 20% than that for GMAW. Both are in the intermediate range for sensitization. It is appropriate to note that the G67 test may not be very indicative for welds because most of the 6.3 x 50-mm sample is parent material, so some of the weight loss reflects parent metal behavior rather than the weld. Based upon these observations, it appears that FSW welds can provide certain advantage in corrosion resistance as well as strength and uniformity across the stir zone.

In order to evaluate the corrosion resistance of plates after 30 years services in equatorial environments, the samples were thermally exposed at 117 °C for four days, which is equivalent to 30 years at 65 °C based on Larson-Miller analysis. [8] The intergranular corrosion resistance were evaluated for samples of the FSW, GMAW and parent metal after such exposure.

The right hand side of Figure 5 provides the G67 weight losses of the FSW and GMAW joints following the high temperature exposure. Both welded samples show significantly larger weight losses than before the high temperature exposure, with the FSW joint exhibiting moderately less weight loss than GMAW joint. It should be aware that the G67 test may not be very indicative for welds because most of the 6.3 x 50-mm sample is parent material, so in this case most of the weight loss for both welds probably reflects mostly the relatively poor behavior of the parent metal behavior rather than much about the welds.

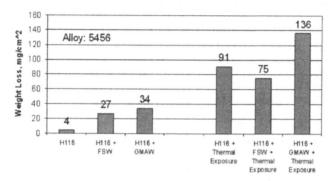

Figure 5: Weight losses by intergranular corrosion testing of joints in 5456-H116 plate

Figure 6 give the comparison of intergranular corrosion attack of samples after ASTM G67 testing. At the microstructural level, the GMAW welds show much more intergranular attack both in the weld zone, further indication that FSW joints provide greater protection from intergranular attack. Overall this evidence supports the likelihood that FSW joining provides better intergranular corrosion for 5xx series Al-Mg alloys.

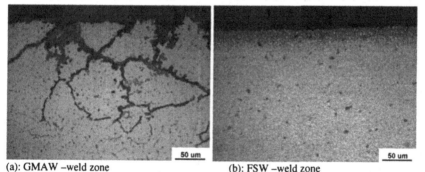

(a): GMAW –weld zone   (b): FSW –weld zone

Figure 6: Comparison of intergranular corrosion attack in G66 testing of FSW and GMAW joints in 6-mm 5456-H116 plate after 4 days thermal exposure at 117^0C (a): GMAW –weld, (b): FSW – weld

Summary

In the research results described herein, the FSW process results in less heat-affected zone and higher strength compared with conventional GMAW joints. The results for this one lot of FSW joints in 6-mm 5456-H116 plate also suggests that FSW joints in Al-Mg alloys have superior resistance to intergranular corrosion attack than GMAW joints for both as-welded and after thermal exposure samples.

Reference

[1]:   William E. Frazier, High-Strength Aluminum Alloys, Advanced Materials & Processes, March 2007, PP 23.

[2]:   Ronan DIF, Tim Warner, and Guy-Michel Raynaud, Corrosion Resistance of Aluminum-Magnisium Alloys for the Marine Market, Proceedings of ICAA-6, Aluminum Alloy, Vol.3, 1998, p1615 – 1620

[3]:   R. S. Mishra, M. W. Mahoney, S. X. McFadden, N. A. Mara and A. K. Mukherjee, Scripta Materialia, 42 (1999) 163-168

[4]:   C. G. Rhodes, M. W. Mahoney, W. H. Bingel, R. A. Spurling and C. C. Bampton, Scripta Materialia, 36(1997), no. 1, 69-75

[5]:   Z. Y. Ma, R. S. Mishra and M. W. Mahoney, Acta Materialia, 50 (2002), no.17, 4419-4430

[6]:   A.P. Reynolds, W.D. Lockwood and T.U. Seidel, Material Sc. Forum, 331-337 (2000), 1719-1724

[7]:   I. Charit, R.S. Mishra and Murray W. Mahoney, Scripta Materialia, 47 (2002), no. 9, 631-636

[8]:   F.R. Larson and James Miller, A Time-Temperature Relationship for Rupture and Creep Stresses, Transactions of the ASME, Vol. 74, ASME, New York, July, 1952, pp 785-771.

[9]:   Kaufman, J. Gilbert; Long, Zhengdong; Ningileri, Shridas. Application of time-temperature-stress parameters to high temperature performance of aluminum alloys. TMS 2007, Feb. 25-Mar. 1, 2007 (2007),   137-146.

Aluminum Alloys: Fabrication, Characterization and Applications II
Edited by: Weimin Yin, Subodh K. Das, and Zhengdong Long
TMS (The Minerals, Metals & Materials Society), 2009

HYDROGEN GENERATION BY ALUMINUM-WATER REACTIONS

P.Rozenak[a] and E. Shani[b]

[a] Hydrogen Energy Batteries LTD, Eshel 44 st. POB 195, Omer 84965, Israel,
[b] Faculty of Health Sciences, Ben-Gurion University of the Negev, Beer Sheva, Israel.

Keywords: Aluminum, Hydrogen, Chemical reaction, Hydride.

Abstract

The aluminum-water surface reaction in the alkaline dissolution of aluminum was studied by secondary ion mass spectroscopy (SIMS) and transmission electron microscopy (TEM). In our experiments, hydrogen (deuterium) absorption in the aluminum surface after reaction with an alkaline solution was characterized. We suggest the interpretation that anodic oxidation during the first anodic scan irreversibly converts the active hydride-covered Al surface to a passive oxide-covered surface. The result suggests that, at least in alkaline solutions, the Al dissolution pathway proceeds through a hydride oxidation step. Hydride may be formed by the etching of Al and hydrogen gas generated cathodically on the surface. The principles of the production of gaseous hydrogen are described.

1. Introduction

In the world today, hydrogen is produced by steam reforming of natural gas. It is also produced by water electrolysis, its most important industrial process. However, the costs of CO_2-free hydrogen production are high and they cannot compete with current prices of oil or natural gas. Therefore, the development of new technologies for hydrogen production, not based on fossil fuels, is becoming increasingly important, in order to provide a clean environment and fuel for life in the 21st century [1,2]. The generation of hydrogen for fuel cell application, by reaction of chemical hydrides with aqueous solutions, reduces storage weight and/or volume, in comparison to high pressure or cryogenic storage [3]. However, hydrogen production from hydrides also has some disadvantages: hydrides are expensive raw materials, considering current hydrogen prices and most of them are unstable and sensitive to air humidity. On the other hand, hydrogen desorption from light metal hydrides, such as alanates, is an endothermic process that requires T≥373K [4]. Recently, certain chemical reactions of aluminum based alloys metals accompanied by hydrogen evolution have received increasing concerns in the field of hydrogen energy because of their potential applications in both hydrogen production and storage. In these reactions, the hydrogen sources such as water are usually used as one of the reactants, from which hydrogen will be extracted with the interaction with high activated metal. This is in the fact an innovative application of an old technology as hydrogen evolution through the displacement reactions of metals was discovered several centuries ago and some of these reactions have already been studies thoroughly [5]. Alkaline solutions are able to corrode the protective oxide layer on the surface of metallic aluminum. Therefore, aluminum is readily dissolved in the alkaline environment at room temperature, resulting in hydrogen production. Of the various alkaline solutions, sodium hydroxide (NaOH) is the one most commonly encountered. Although it is a well-know reaction, which indeed provides compact gaseous hydrogen, only a few studies have been published directly investigate the production of surface hydride formation during the dissolution of aluminum [6]. Dissolution of aluminum in NaOH solutions results in substantial rates on hydrogen absorption into the metal [7], as well as the formation of nanoscale voids or hydrogen bubbles in the near surface region of the metals [8-11].

The presence of hydride on the dissolving aluminum surface leads us to propose a chemical mechanism reaction for alkaline corrosion of aluminum. Also, the direct participation of hydrogen in the dissolution reaction provides a possible mechanism by which hydrogen defects can be formed and may help explain the observed high rates of hydrogen generation and adsorption related void (bubble) formation during the dissolution process.

2. Experimental results and discussion

Aluminum oxide or alumina, Al_2O_3, occurs in various forms on metal aluminum. The physical and chemical properties of alumina depend, to a large extent, on the temperature, time, chemical and others conditions, during its preparation. Actually, the corrosion behavior of aluminum is determined essentially by the behavior of the oxide film with which it is almost always covered towards the corroding media to be considered. Cases of bad resistance to corrosion are often connected with a change in this oxide film, notably in its degree of hydration and porosivity. Chemical potential-pH equilibrium diagram for the system aluminum-water at $25°$ considering as with a change in this oxide film, notably in its degree of hydration and porosivity, can be seen in Fig.1. Aluminum is seems to be a very base metal, as the whole its domain of stability lies below that of water. In the presence of sufficiently acid solution, its decomposes water with the evolution of hydrogen, dissolving as trivalent Al^{+++} ions; monovalent Al^+ ions, and as aluminate ions AlO^-_2 and leaving the electrons on the metal. In non-complexion solutions of pH roughly between 4 and 9, aluminum tends to become covered with a film of oxide. The most stable form of $Al_2O_{3,}$ is trihydrited aluminum oxide, hydragillite, with monoclinic structure as can be seen from the diagram. The diagram not taken into account the cations $Al(OH)^{++}$ and $Al(OH)_2^+$ (or AlO^+) which seem to relate only to chloride complexes. The equilibrium diagram is valid only in the absence of substances with which aluminum can form soluble complexes or insoluble salts. The principal aluminum complexes are those formed with the anions of organic compounds (acetic, citric, tartaric, oxalic ions etc.) and also the fluorine complexes and sparingly soluble salts, the phosphate and oxinate may be mentioned. Calcium aluminate is also sparingly soluble. When alkali is added to an acid solution, an aluminate precipitate is obtained, which is a hydroxide gel, corresponding practically to the composition $Al(OH)_3$ and amphoteric in nature. However, this aluminum hydroxide gel is not stable and it crystallizes in the course of time, to give, first, the monohydrate γ-$Al_2O_3 \times H_2O$ (boehmite), crystallizing in the rhombohedral system, then the trihydrate $Al_2O_3 \times 3H_2O$ (bayerite), crystallizing in the monoclinic system. This development of aluminum hydroxide is known as "aging". In acid or alkaline solution, the aluminum will be attacked as soon as the oxide film is eliminated. This dissolution is slower in acidic than in alkaline solutions [7]. On the metal-oxide interface in aluminum, atomic hydrogen combines to form molecular hydrogen in the porous hydroxide layer [8]. This process effectively makes the interface coherent and, as a result, surface film blistering (surface bubbles) is initiated. The hydrogen gas pressurizes the surface bubbles and extra material for the surface bubble wall is provided by the low density porous hydrous oxide. The compact barrier layer cannot be penetrated by the gas molecules and hence, a high pressure of gas can develop within the surface bubbles (blisters) [8]. However, many works have been published dealing with aluminum hydride formations [13-19]. Secondary ion mass spectrometry is a highly sensitive surface analytical method that has been used extensively for characterization of deuterium (hydrogen form in air from humidity) absorption into pure aluminum [7]. In the present study, the deuterium distribution through the specimen thickness, during the chemical charging at room temperature, was characterized. Fig.2. SIMS profiles of $^{16}O^-$, $^{27}Al^-$, $^1H^-$ and $^2D^-$ vs. depth in a single crystal of aluminum chemically charged for 1h in 0.1N NaOD solution at room temperature and a crater profile done by profilometer. A number of D depth profiles were obtained for each sample. Typical results are shown for distribution of $^-O^{16}$, $^-Al^{27}$, $^-H^1$ and $^-D^2$ in single crystals of Al chemically charged for 2h at room temperature. Ions of oxygen $^-O^{16}$ (upper curve) showed a slight change of value to the depth under the specimen surface. Ions of $^-Al^{27}$ exhibited maximum values on the sample surface and then decreased exponentially to a depth of 0.1 μm. The aluminum samples exhibited ions of hydrogen $^1H^-$ during the SIMS measurements. The hydrogen exhibited maximum values at the samples' surface, which decreased linearly to a depth of the 0.4 μm and then decreased to about 0.5 μm in the low hydrogen penetration zones. Hydrogen can be formed from moisture in the air [20] and diffused into the specimens during ion irradiation of high purity aluminum [21]. Chemical charging of a single crystal of aluminum for 2h at room temperature exhibited the formation of a high concentration of deuterium on the surface of the samples, decreased linearly to a depth of 0.1 μm and then decreased to about 0.5 μm in the low deuterium penetration zones. Aluminum hydroxide $Al(OH)_3$ and $Al(OD)_3$ layers of a thickness of about 0.08 μm were formed on the surface layer regions. Aqueous solutions of NaOH are known to be aggressive toward aluminum and chemical charging at pH 12 resulted in dissolution of the Al, as well as introduction of hydrogen at concentrations of 600-1300 appm [5]. It is suggested that the introduction of solute hydrogen results from the reaction of Al with H_2O and the alkaline solution serves to remove the Al_2O_3 allowing the reaction between the Al and H_2O. In this study and Adhikari et.al work [6], indicated that alkaline dissolution of Al occurs by the continuous anodic oxidation of hydride formed by the accompanying cathodic reaction. It is plausible that hydride forms by reaction of cathodically generated hydrogen with the aluminum surface, a process which has been found to occur during hydrogen exposure of Al surfaces. Aluminum corrosion mechanisms involving hydride have been proposed previously, but the study is the first analytical detection of hydride formed by dissolution. Continuous hydride formation and oxidation reaction, suggests that the dissolving aluminum metal surface is the first analytical detection of hydride formed by dissolution. Continuous hydride formation and oxidation reaction, suggests that the dissolving aluminum metal surface is not covered by a resistive oxide, as previously considered. Instead, it may be covered by a hydride layer, which prevents oxidation of surface of Al atoms. The direct participation of hydrogen in the dissolution reaction provides a possible avenue by which hydrogen-defects can be continuously formed and may help explain the observed high rates of hydrogen forms and absorption and void (bubble) formation during dissolution [5]. The formation and oxidation reaction, suggests that the dissolving aluminum metal surface is not covered by a resistive oxide, as previously considered. Instead, it may be covered by a hydride layer, which prevents oxidation of surface of Al atoms. The direct participation of hydrogen in the dissolution reaction provides a possible avenue by which hydrogen-defects can be continuously formed and may help explain the observed high rates of hydrogen forms and absorption and void (bubble) formation during dissolution [5]. The generation of gas bubbles in the solution on the aluminum surface, also observed in these experiments, was detected immediately after the aluminum was mixed in the reactor with various NaOH aqueous solutions [2, 22-25]. The generated gas was identified as hydrogen of

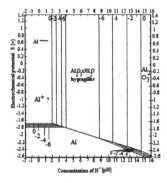

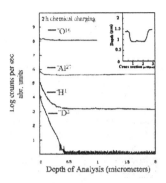

Fig.1. Electrochemical potential versus concentration of H⁺ diagram for the system aluminum-water, at 25°C.

Fig.2. SIMS profiles of $^{16}O^-$, $^{27}Al^-$, $^{1}H^-$ and $^{2}D^-$ vs. deep in the single crystal of aluminum, chemically charged for 2 hour in 0.1 N NaOH solution at room temperature.

high purity [5]. As shown in Fig. 3, the generation rate of H_2 gas R_{H_2}, increased with the chemical reaction time up to 600 seconds, beyond which, it started to decrease. With the gas generation, the temperature of the NaOH aqueous solution rose quickly with the reaction time and reached a maximum of 338K within 800 seconds after the reaction terminated. On the other hand, the pH of the solution gradually decreased with the reaction. When the reaction time exceeded 1200 seconds, the temperature of the NaOH solution began to decrease. However, the pH of the solution showed a gradual increase beyond this point. The pH increase was found to be preceded by precipitation that occurred at t~1000 seconds and the precipitation compound was later identified as containing aluminum hydroxides. The generation of gas bubbles of volume V_{H_2} was detected immediately after the aluminum powder and NaOH aqueous solution were mixed in the reactor. The bubble diameters, densities and distribution ranged in size from small (some millimeters in diameter) in the initiation times of the reaction, to very large (decimeters in diameter), in the longer times. Accumulated volume V_{H_2} increased exponentially to that of 7 dm³ after 800 seconds and continued to increase slightly with the reaction time, reaching a maximum of 8.75dm³ within 2000 seconds after the reaction terminate. The essence of this method is that the aluminum and the water based alkaline electrolyte are brought into contact as needed, to create a reaction for the production of hydrogen. The purpose of using the alkaline solution is to dissolve the thin protective layer of oxidation that forms as soon as the metal is exposed to the air and to provide the water with access to a clean aluminum surface. The essence of this method is that the aluminum and the water based alkaline electrolyte are brought into contact as needed, to create a reaction for the production of

180

hydrogen and then to supply it directly to a fuel cell power system [26-27]. The purpose of using the alkaline solution is to dissolve the thin protective layer of oxidation that forms as soon as the metal is exposed to the air and to provide the water with access to a clean aluminum surface. The basic exothermic is typical of the chemical reaction in (1). To produce 1 kg of hydrogen in this reaction requires: 18 kg of water (as the primary source of hydrogen), 9 kg of aluminum (as the consumable catalyst of the chemical reaction of the water split) 2 kg of a non-consumable chemical mix (sodium hydroxide plus a proprietary mix of additives, needed to maintain the effective alkaline solution). In the reaction with oxygen inside the fuel cell power plant, 1 kg of hydrogen, generates 9 kg of water (as a fuel cell by-product). By recycling this water to the hydrogen generator, only 9 extra kg of water will be required for the next cycle of hydrogen production. This technique can increase the yield to over 7% of the components' weight aluminum hydroxide, Al(OH)$_3$, produced as a byproduct, can be fully recycled to aluminum and can be reused for hydrogen production an unlimited number of times. Unique additives allow the electrolyte solution to react with the aluminum (no alloys) plates to form aluminum hydroxide in crystalline particle form, which falls to the bottom of the chemical reactor container, leaving the reaction surface clear and active.

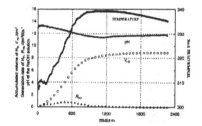

Fig.3. Time evolution of pH, accumulated hydrogen volume V_{H_2} and generation rate R_H of hydrogen obtained from aluminum powder in NaOH solution [22].

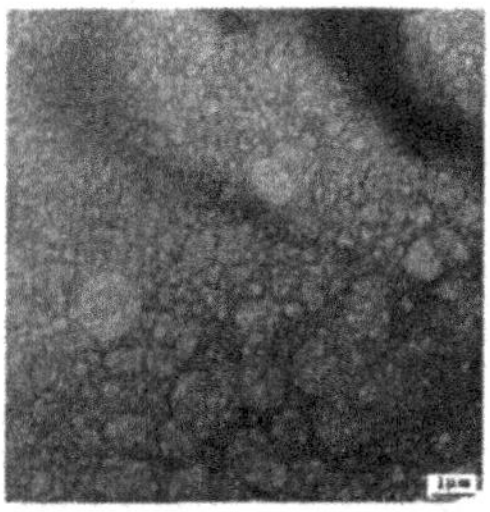

Fig.4. TEM micrograph showing surface bubbles formation in chemically charged aluminum specimen during 24 h at room temperature.

3. Conclusions

Secondary ion mass spectrometer is highly sensitive for surface analytical method that has been used extensively for characterization of H absorption into the aluminum. We suggest the interpretation that anodic oxidation during the first anodic scan irreversibly converts the active hydridecovered Al surface to the passive oxide-covered surface. In alkaline solutions, the Al dissolution pathway proceeds through a hydride oxidation step. Hydride may be formed by the etching of Al and generated hydrogen gas cathodically on the surface.

4. References

(1) A. Bauen, J. Power Sources, 2006; 157: 893-901. (2) L. Soler ,J. Macanás , M. Muñoz , J. Casado , Proceedings International Hydrogen Energy Congress and Exhibition IHEC, 2005; Istanbul Turkey, 13-15 July. (3) U. Eberle , G. Arnold , R von Helmot, J. of Power Sources, 2006;157: 456-60. (4) Z. F. Hou , J. of Power Sources, 2006; 159: 111-15.

(5) Z. H.Wang, D. Y. C Leung , M. K. H Leung , M. Ni , Renewable & Sustainable Energy Reviews, 2008, In press. (6) S. Adhikari , J. Lee , K. R. Hebert, Journal of The Electrochemical Society, 2008; 155 (1); C16-C21.

(7) P. Rozenak, B. Ladna B, H. K. Birnbaum , J. of Alloys and Compounds, 2006; 415: 134-42.

(8) H. K. Birnbaum, C. Buckley , F. Zeides , E. Sirois , P. Rozenak , S. Spooner , J. S. Lin, J. of Alloys and Compounds, 1997; 253-254: 260-4; P. Rozenak , E. Sirois E, B. Ladna , H. K. Birnbaum and S. Spooner, Characterization of hydrogen defects forming during chemical charging in the aluminum. J. of Alloys and Compounds, 2005; 387: 201-10.

(9) P. Rozenak , J. of Alloys and Compounds, 2005; 400: 106-11.

(10) P. Rozenak, Int. J. Hydrogen Energy, 2007; 32: 2816-23.

(11) R. Huang,, R.K. Hebert, L.S. Chumbley, Journal of The Electrochemical Society, 151 (7), 2004: B379-B386. (12) M.Poubaix, Atlas of Electrochemical Equilibria in Aqueous Solutions, Pergamon Press, 1966.

(13) M. Appel, J. P. Frankel, J. of Chemical Physics, 1965; 42,11: 3984-90.

(14) P. J. Herley, O. Christofferson ,J.A. Todd, J. of Solid State Chemistry, 1980; 35: 391-401.

(15) P. J. Herley, O. Christofferson, J. Phys. Chemistry, 1981;85: 1887- 92.

(16) P. J. Herley,O. Christofferson, R. Irwin, J. Phys. Chem., 1981; 85: 1874-81.

(17) P. J. Herley, O. Christofferson, J. Phys. Chem., 1981; 85: 1882-86.

(18) P. J. Herley, J.A. Todd, Letters, J. of Mater. Science, 1982; 1 ;163.

(19) P. Breisacher, B. Siegel, J. Am. Chem. Soc., 1983; 105: 1704-5.

(20) M. R. Louthan JR, A. H Dexter , Metal.Trans, 1975; 6A; 1655-7.

(21) S. Foruno ,K. Izuik, K. Ono,T. Kino , J. of Nuclear Materials, 1985; 133-134: 400-4.

(22) T. Hiraki, M. Takeuchi, M Hisa, T Akiyama, vol.46, no.5 (2005) 1052.

(23) S.S. Martinez, W.L. Benites, A.A.A Gallegos, P.J. Sebastian, Solar Energy Materials & Solar Cells, 2005; 33 : 237-43. (24) L. Soler, J. Macanas, M. unos,J. Casado, Int. J. of Hydrogen Energy, 2007; 32: 4702-4710.

(25) D. Belitskus, J.of Electrochemical Society, 1962; 116, 8 : 1097-99.

(26) E.D. Wang, P.F. Shi, C.Y. Du, X.R. Wang, A mini-type hydrogen generator from aluminum for proton exchange membrane fuel cell, Journal of Power sources, 2008, In press.

(27) S.S. Martinez, L.A. Sanchez, A.A.A. Gallegos, P.J. Sebastian, Int. J. of Hydrogen Energy, 32, 2007; 3159-3162.

Aluminum Alloys: Fabrication, Characterization and Applications II
Edited by: Weimin Yin, Subodh K. Das, and Zhengdong Long
TMS (The Minerals, Metals & Materials Society), 2009

Splitting Water with Al Rich Alloys: Structure and Reaction Kinetics

Jeffrey T. Ziebarth[1], Jerry Woodall[1], Go Choi[1], C.R. Allen[1]; J-H Jeon[1]; Deborah Sherman[1]; Robert Kramer[2]

[1]Purdue University, West Lafayette, IN 47907 USA
[2]Purdue University, Hammond, IN 46323 USA

Keywords: Aluminum, Gallium, Water, Hydrogen

Abstract

Solid alloys of aluminum, gallium, indium and tin are capable of reacting with water at room temperature to form hydrogen, alumina, and heat. The alloys are shown to contain a phase of solid aluminum-rich grains with small amounts of gallium. In and Sn have nearly zero solid solubility in Al, and energy dispersive x-ray (EDX) analysis found In and Sn to be in the grain boundaries together with Al and Ga. It is believed that the grain boundary phase becomes liquid at or near room temperature and as a result enables reaction with water. When these alloys react with water or oxidize in air, EDX results show that the reaction occurs at or near the grain boundary. Current research efforts focus on studying the reaction mechanisms for alloy compositions containing 50 wt% and 95 wt% Al, with 95 wt% Al alloys being the more interesting of the two from an economics standpoint.

Introduction

The stoichiometric equation for the reaction of aluminum with water is given in Equation 1 with the enthalpy value calculated using the standard molar enthalpy (heat) of formation at 298.15 K for each chemical [5].

$$Al + \tfrac{3}{2}H_2O \longrightarrow \tfrac{1}{2}Al_2O_3 + \tfrac{3}{2}H_2 + 409.15 \text{ kJ/mol} \tag{1}$$

Burning the hydrogen produced from reacting 1 mole of aluminum with water will produce about 428.7 kJ, as shown in Equation 2.

$$\tfrac{3}{2}H_2 + \tfrac{3}{4}O_2 \longrightarrow \tfrac{3}{2}H_2O + 428.7 \text{ kJ/mol} \tag{2}$$

Thus there is approximately 837.85 kJ of usable energy in every mole of aluminum. Alloying aluminum with gallium, indium, and tin provides access to this energy by disrupting the means by which aluminum self-passivates. Assuming the total energy content of aluminum to be as calculated above, a typical 95 wt% aluminum alloy can have an total energy density as high as 8.18 kWh/kg. Recycling the resultant oxide material by fused-salt electrolysis has an energy cost of about 13 DC kWh/kg Al [8], and enables a renewable cycle for delivering hydrogen and heat energy to remote applications. Two compositions to date have been put to experimental analysis: 95 wt% Al, 3.4 wt% Ga, 1.1 wt% In, 0.5 wt% Sn, and 50 wt% Al, 34 wt% Ga, 11 wt% In, 5 wt% Sn. In both compositions, the mass ratio of Ga:In:Sn is kept constant.

Quaternary System Hypothesis for Non-equilibrium Cooling

There is currently no literature on the nature of this quaternary system. As a brief overview, the binary and ternary equilibrium alloy phase diagrams for these elements are discussed so as to suggest plausible characteristics of the quaternary system. The bulk of the current work originated on the simple Al-Ga alloy phase diagram [2],[4],[7]. One can conclude from this phase diagram that a solid phase of aluminum can exist with a certain amount of gallium in it, up to about 20 wt% [3]. However, at equilibrium, solid phases of aluminum cannot contain any amount of indium or tin, as evidenced by the binary alloy phase diagrams for Al-In and Al-Sn [3].

It is important to note, however, that rapid cooling was used in the fabrication of these alloys, and hence, non-equilibrium arrangements of these metals are likely. For compositions of less than 20 wt% gallium, it is possible that not all of the gallium is quenched into the solid aluminum phase and exists in a separate phase containing aluminum, indium and tin. Results from EDX imaging tend to support this claim. The percentage of gallium not quenched into the solid phase is likely to be related to the cooling rate. The gallium not in the solid aluminum phase has the potential to form a eutectic with the indium and tin components of the alloy in the grain boundaries. The minimum freezing point of this ternary system is reported to be 10.7°C [1],[6]. It is hypothesized that the resulting quaternary alloy system consists of at least two phases: one solid phase containing Al-Ga, and one liquid phase of Ga-In-Sn containing dissolved Al. This hypothesis is supported by experimental evidence showing that Al-Ga-In-Sn alloys react at room temperature, where as Al-Ga alloys require elevated temperatures for reaction [2],[4],[7]. This would not be the first instance of liquid phases present in aluminum alloys. Impurities of bismuth and lead are responsible for a phenomenon known as "hot cracking" because of their low solubility in aluminum. Under temperature conditions sufficient to liquefy these lead or bismuth phases, the aluminum material becomes prone to failure [3]. The key difference here, however, is that these liquid phases would take place below the melting point of pure gallium.

Microscopy

Samples of 95 wt% aluminum alloy made in a furnace using commercial grade aluminum were analyzed using EDX to determine the distribution of the alloy's constituent elements within its microstructure. The target composition of this alloy was 95 wt% Al, 3.4 wt% Ga, 1.1 wt% In and 0.5 wt% Sn. Samples were polished using an alumina film and stored under nitrogen to minimize oxidation until being placed in the EDX sample holder.

At first glance, the distribution of elements in the sample indicate the sample is non-homogeneous, and that phase separation has occurred. This can be seen in Figure 1. Brighter regions are indicative of heaver elements, darker regions indicate lighter elements. Long, straight lines are the result of scratches from polishing. The brighter regions appear to form a network through the sample, which most likely indicate the presence of grain boundaries containing indium and tin. The black regions indicate areas where aluminum has reacted with oxygen to form an oxide compound.

Again, in Figure 2, segregation of phases can be clearly seen. Spectra from four regions were analyzed to determine the compositions of each. The spectrum from the first image

Figure 1: 95 wt% Al alloy sample with segregation of In and Sn

shows that bright regions are rich in indium and tin, as can be seen in Table 1. The atomic percentages indicate that some of the aluminum in the region has been oxidized. The spec-

Element	App Conc.	Intensity Corrn.	Weight %	Weight % Sigma	Atomic %
O	17.27	0.3633	21.18	0.55	38.18
Al	67.67	0.5992	50.31	0.42	53.78
Ga	11.81	0.9139	5.76	0.14	2.38
In	32.61	0.8668	16.72	0.26	4.20
Sn	10.85	0.8016	6.03	0.25	1.46
Totals			100.00		100.00

Table 1: Spectrum 1 results

trum of region 2 seen in Table 2 is in stark contrast to that of region 1 as is expected based on the image in Figure 2. This region is marked by its complete absence of indium and tin. The lower fraction of oxidized aluminum is of interest. The spectrum analysis of region 3 shown in Table 3 shows similarities to region 1, however there is a considerably higher content of indium. This suggests that even the individual phases are non-homogeneous. Also of interest is the roughly 3:1 mole ratio of oxygen to aluminum. This is unexpected, since for pure Al_2O_3 the mole ratio should be 3:2. The source of this unexpected ratio is still being investigated. Region 4 appears to have a faint streak running vertically through it. Its spectral analysis is given in Table 4. The analysis indicates a low portion of oxygen and indium, suggesting that indium, and not just gallium alone, plays a role in the potential for the alloy to react with oxygen. Certain compositions of gallium, indium and tin form a eutectic mixture which is liquid at room temperature[6] and it is hypothesized that this eutectic exists as a liquid phase in the alloy along side other solid phases of aluminum and

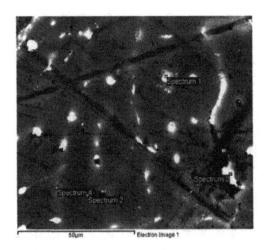

Figure 2: 95 wt% Al alloy sample four regions

Element	App Conc.	Intensity Corrn.	Weight %	Weight % Sigma	Atomic %
O	10.33	0.5163	12.91	0.41	20.38
Al	130.95	1.0084	83.76	0.41	78.41
Ga	4.61	0.8914	3.34	0.13	1.21
Totals			100.00		100.00

Table 2: Spectrum 2 results

gallium. This is plausible, because unlike gallium, indium and tin are not soluble in solid aluminum as predicted by their respective equilibrium binary alloy phase diagrams [3].

Conclusions

It is not fully understood how the alloys split water in the solid phase. By examining the microstructure of the alloy before and after exposure to water, it is hoped that the processes governing the reaction will be revealed. Determining how the microstructure oxidizes will indicate how the components of the alloy (grains and grain-boundaries) participate in the reaction and suggest a model for the reaction's governing mechanisms. The ability of the alloy to react at a specific site seems to show a correlation with the presence of gallium and indium at that site. This would suggest that liquid phases must be present at room temperature for a significant reaction to take place. Earlier alloys made without indium and tin required heated water to begin reacting and therefore support this claim. Future work will center around using other tools such as differential scanning calorimetry (DSC) to determine the presence of low temperature solid-to-liquid phase transitions.

Element	App Conc.	Intensity Corrn.	Weight %	Weight % Sigma	Atomic %
O	56.09	0.4118	46.40	0.46	70.37
Al	40.41	0.5263	26.16	0.26	23.52
Ga	6.59	0.8912	2.52	0.10	0.88
In	54.94	0.9438	19.83	0.27	4.19
Sn	12.85	0.8616	5.08	0.24	1.04
Totals			100.00		100.00

Table 3: Spectrum 3 results

Element	App Conc.	Intensity Corrn.	Weight %	Weight % Sigma	Atomic %
O	4.00	0.4795	5.46	0.49	9.33
Al	122.17	0.9243	86.59	0.49	87.67
Ga	10.02	0.9041	7.26	0.16	2.84
Sn	0.74	0.7047	0.69	0.13	0.16
Totals			100.00		100.00

Table 4: Spectrum 4 results

References

[1] D.S. Evans and A. Prince, Ga-In-Sn Phase Diagram, ASM Alloy Phase Diagrams Center, P. Villars, editor-in-chief; H. Okamoto and K. Cenzual, section editors; http://www.asminternational.org/AsmEnterprise/APD, ASM International, Materials Park, OH, 2006.

[2] Jerry .M. Woodall, Jeffrey Ziebarth, and Charles R. Allen "The Science and Technology of Al-Ga Alloys as a Material for Energy Storage, Transport, and Splitting Water," Proc. 2nd Energy Nanotechnology International Conference, Sept 5-7, 2007 Santa Clara, CA.

[3] Hugh Baker, editor. *ASM Handbook, Vol. 03: Alloy Phase Diagrams.* ASM International, 1992.

[4] J.M. Woodall, Jeffrey T. Ziebarth, Charles R. Allen, J. Jeon, G. Choi, and Robert Kramer, "Generating Hydrogen on Demand by Splitting Water with Al Rich Alloys", Clean Technology 2008, Boston, Massachusetts, June 1-5, 2008.

[5] David R. Lide, editor. *CRC Handbook of Chemistry and Physics.* CRC Press/Taylor and Francis, 88th edition edition, 2007. see table: "Standard Thermodynamic Properties of Chemical Substances".

[6] C. Lu and K.T. McDonald. Low melting temperature metals for possible use as primary targets at a muon collider source. Technical report, Princeton University, May 1997. updated June 5, 1998; added estimate of onset of thermal shock; June 12, added phase diagrams for Ga-Sn, Ga-Zn and Ga-In-Sn.

[7] J. M. Woodall, Jeffrey T. Ziebarth, Charles R. Allen, Debra M. Sherman, J. Jeon, G. Choi, "Recent Results On Splitting Water with Aluminum Alloys", Materials Innovations in an Emerging Hydrogen Economy conference (Hydrogen 2008), Cocoa Beach, Florida, Feb. 24-27, 2008.

[8] Barry J. Welch. Aluminum production paths in the new millennium. *Journal of the Minerals, Metals and Materials Society*, 51(5):24–28, May 1999.

Aluminum Alloys:
Fabrication, Characterization and Applications II

Composite and Foam

Session Chair

Zhengdong Long

Aluminum Alloys: Fabrication, Characterization and Applications II
Edited by: Weimin Yin, Subodh K. Das, and Zhengdong Long
TMS (The Minerals, Metals & Materials Society), 2009

JOINING OF ALUMINUM 5754 ALLOY TO CARBON FIBER REINFORCED POLYMERS (CFRP) BY ULTRASONIC WELDING

Frank Balle, Guntram Wagner, and Dietmar Eifler

Institute of Materials Science and Engineering;
University of Kaiserslautern; P.O. Box 3049, 67653 Kaiserslautern, Germany

Keywords: Ultrasonic welding, Joining, Aluminum, CFRP, Lightweight, Multi-Material-Design

Abstract

Ultrasonic metal welds were performed to realize aluminum alloy/carbon fiber reinforced polymer (CFRP) – joints. Important advantages of ultrasonic welding are short welding times and welding temperatures below 450°C. Important steps of the process are the softening and displacing of the polymer out of the welding zone by the ultrasonic shear oscillation. In the following, in contrast to conventional plastic welding processes, a direct contact between the aluminum surface and the carbon fibers takes place. The bonding mechanisms can be shown in detail by scanning electron microscopy. In first welding tests shear strengths of about 30 MPa were realized for AA5754/CF-PA66 – joints. By using stepwise profiles for the welding force the tensile shear strength could be increased up to 45 MPa. Special surface pre-treatments of the metal, like shot peening or etching, lead also to 60% higher joint strengths. Furthermore, the influence of aging in selected climates was investigated.

Introduction

The current demand for lightweight constructions leads to an increasing application of materials like aluminum, magnesium and fiber reinforced polymers and their combinations. Hence the predominant aim of innovative products in the automotive and aircraft industry but also in railway transportation and engineering in general is the reduction of the weight of the components. For the development of new lightweight products detailed knowledge of monotonic and cyclic deformation behavior of joints made of dissimilar materials is obligatory. To fulfill these challenges appropriate joining techniques are necessary.

The ultrasonic welding of similar materials e.g. wires for cable harnesses or plastics for commercial packaging is already established in industrial manufacturing. However, a central purpose of the research at the Institute of Materials Science and Engineering (WKK) at the University of Kaiserslautern, Germany, is to develop new application fields for the ultrasonic metal welding process, e.g. the joining of dissimilar materials such as metal with glasses or fiber reinforced polymers (FRP) [1].

In comparison to other joining techniques like adhesive bonding or brazing, ultrasonic welding is characterized by a low energy input and consequently low temperatures in the welding zone as well as short welding times [2]. Until now ultrasonic plastic welding is typically used for joining FRP to each other, but in the case of metal/FRP-joints the ultrasonic plastic welding method only enables a joining between the metal and the matrix and not with the load bearing fibers of the FRP [3]. Recent investigations at the WKK show that ultrasonic metal welding is a suitable alternative to join CFRP with sheet metals like aluminum alloys or pure aluminum plated steel [4, 8]. In the following the ultrasonic metal welding technology and the mechanical properties of the realized joints are presented.

Ultrasonic Welding Process

Recent results show, that beside compact glass also glass fiber textiles or carbon fiber textiles with or even without thermoplastic matrix can be welded to metals by ultrasonic metal welding [3, 7, 8]. For the investigations described in this paper ultrasonic spot metal welding systems were used for joints between sheet metals (AA5754) and carbon fiber reinforced thermoplastic composites (CFRP) with PA66 or PEEK matrix. One advantage of the ultrasonic oscillation parallel to the surface of the joining partners, typical for ultrasonic metal welding, is the possibility to realize a direct contact between the metal and the load bearing fibers of the reinforced composite without damaging the fibers by the welding process. Scanning electron micrographs (SEM) of the welding zones show, that the ultrasonic metal welding process removes the matrix between the fiber reinforcement and the metal whereby the metallic surface gets into contact to the fibers [4, 8]. In comparison to ultrasonic plastic welding 60% higher tensile shear strength can be achieved by using ultrasonic metal welding [3]. The main components and the operating mode of a spot welding system are described in Figure 1.

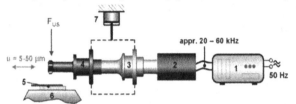

Figure 1. Principle of ultrasonic spot metal welding

The main components are an ultrasonic generator (1), a converter (2), a booster (3) and a welding tool called sonotrode (4). The two materials (5) to be welded are pressed on an anvil (6) under a static pressure perpendicular to the welding zone. This clamping force is one important process parameter. The ultrasonic generator converts the 50 Hz main voltage into a high frequency alternating voltage output of 20 kHz. In the converter this oscillation is transformed into mechanical oscillations of the same frequency due to the inverse piezoelectric effect.

The necessary oscillation amplitude in the welding zone, the second important process parameter, the welding amplitude, is ensured by an appropriate design of the booster and the sonotrode. The amplitude typically ranges between 5 µm and 50 µm. Together with the third process parameter, the specific welding energy input, the welding process can be described completely. Moreover, there are several material parameters like surface structure or sheet thickness, which have an influence on the achievable mechanical properties of the joints [1, 8].

Experimental Setup

In the following selected results for welds between the aluminum alloy 5754 and two different CFRP sheets are presented. For the aluminum sheet a thickness of 1 mm and for the CFRP 2 mm were chosen. The fiber reinforcement of the CFRP is a C-textile Satin 5H-fabric with a weight per unit area of 285 g/m². The fiber volume fraction of the 2 mm thick organic sheets is about 48%. It was manufactured in an autoclave process by using six layers of CF-fabric. The welding area of the sonotrode is 10 x 10 mm². Since it is not possible to determine the exact geometry of the joining area the shear strength is calculated by the ratio of the achieved tensile shear force related to the nominal sonotrode contact area. Besides a high reproducible clamping of the specimens on the anvil, it is necessary to control the welding force during the joining process accurately by an integrated force measuring device. Therefore a special clamping system was developed at the WKK, see Figure 2a). The specimen geometry is shown in Figure 2b).

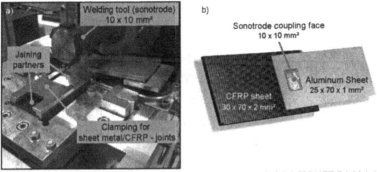

Figure 2. a) Ultrasonic spot welding system, b) Ultrasonic welded AA5754/CF-PA66-joint

A statistical model named "central composite design circumscribed (CCC)" was used to investigate the weldability of the Al/CFRP-joints. In comparison to a stepwise variation of each welding parameter, this model for non-linear relationships allows to find the optimum parameters with considerably less welding steps. An important advantage of the CCC-model is the description of the mutual dependence of the three central welding parameters oscillation amplitude, welding force and energy in relation to the achievable tensile shear strength of the joints [5]. The design of the CCC-model is based on the three significant process parameters mentioned above with five different settings for each. The suitable ranges of the process parameters, which allow high strength joints, were estimated in preliminary investigations. For joints of AA5754 and CF-PA66 composites appropriate welding parameters for the force F_{US} ranges between 100 N and 220 N, for the amplitude u between 37 µm up to 43 µm and for the energy W_{US} between 1700 Ws and 2300 Ws. The used CCC-model leads to 18 different parameter triples and allows to reduce the necessary welds of a factor of 7 in comparison to a conventional stepwise test procedure. Simultaneously the reproducibility of the joint strength is improved. All combinations of the welding parameters were proved in tensile shear tests. For each parameter combination twelve welds were performed.

Results

Figure 3 shows selected results of the investigations based on the CCC-model. On the left hand side the evaluation of the average tensile shear strength for a constant welding energy of $W_{US} = 2160$ Ws is presented. A peak value can be observed for an oscillation amplitude of about 40 µm and a welding force of $F_{US} = 140$ N. Two-dimensional cuts of the diagram are necessary to define the optimum welding parameters exactly. For example on the right hand side of Figure 3 the progression of the tensile shear strength for a suitable welding force (F_{US}) of 140 N and an oscillation amplitude (u) of 40 µm is presented. The welding energy (W_{US}) is varied between 1800 Ws and 2200 Ws. Beside the course of the average tensile shear strength the lower and upper confidential interval for 95% is specified.

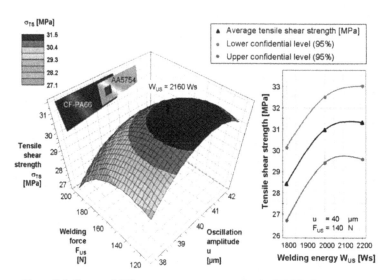

Figure 3. Influence of different process parameters for AA5754/CF-PA66-joints

An increasing welding energy leads up to 2160 Ws to a progressive displacement of the CFRP matrix and a more intensive contact between the metal sheet and the fibers. As a result higher tensile shear strength values are achieved. After the maximum value as a consequence of a too high energy input a damage of the textile or the metal sheet occurs and causes a decrease of the tensile shear strength follows. The courses of the welding force and the oscillation amplitude show the same tendencies. Taking into account all three parameter variations, maximum tensile shear strength of 31.5 MPa can be determined.

The welding time for the evaluated parameters lies approximately at 3.5 s. The welding temperatures were measured by micro-thermocouples which have been integrated in the composites before welding. The maximum temperature, measured in the AA5754/CF-PA66 welding area, was 385°C.

It is also possible to weld pure aluminum and Al-plated unalloyed steel with carbon fiber reinforced polyamide (CF-PA66). For AA1050/CF-PA66-joints appropriate welding parameters lead to tensile shear forces of 2460 N corresponding to a tensile shear strength of about 25 MPa. In this case the tensile strength of aluminum sheet (AA1050) limits the joint strength [4]. For welds performed with Al-plated unalloyed steel and CF-PA66 a tensile shear strength of about 31 MPa was realized.

In addition, the weldability of the high performance composite CF-PEEK with sheet metals was investigated successfully. For these combinations up to now shear strengths of about 30 MPa (AA5754/CF-PEEK) were achieved. A further increase of the joint strength can be obtained by implementing a stepwise decrease of the welding force during the welding process (Figure 4).

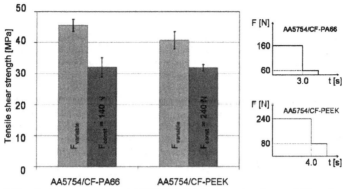

Figure 4. Tensile shear strength for AA5754/CFRP-joints by using time-variable and pressure-variable sequences of the welding force

In the case of different but constant welding forces during the welding process the tensile shear strengths are on the same level with about 32 MPa. But for appropriate force profiles for both material combinations (right side of Figure 4) the achievable tensile shear strength increases substantially. For AA5754/CF-PA66-joints the value improves considerably about 40%. In relation to PA 66 with approx. 260°C the melting temperature of 335°C for PEEK is substantially higher [6]. Hence the welding forces as well as the welding energy were increased from 160 N to 240 N and from 2160 Ws to 2840 Ws, respectively, to attain the necessary displacement of the matrix. As a result the welding time for the AA5754/CF-PEEK-joints increases about 1.5 s up to 5 s and the maximum process temperatures to 450°C. By using mechanical or chemical surface pre-treatments of the sheet metal the weld strength can be increased up to 54 MPa and also the long-term stability can be positively influenced (Figure 5).

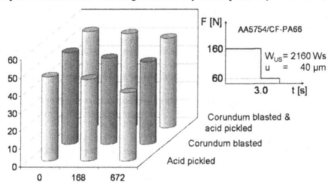

Figure 5. Aging of surface-pretreated AA5754/CFRP-joints (T = 40°C, humidity = 95%)

The ultrasonic spot welded joints were tested after aging of one and four weeks in an ambient temperature of T = 40°C and a humidity of 50%. For a mechanical and a combined pre-treatment of the aluminum sheets (corundum blasting and pickling in nitric acid) nearly no decrease of the tensile shear strength was determined.

195

Conclusions and Outlook

The ultrasonic metal welding technique was applied successfully to join metal sheets with CFRP for the first time. By using the CCC-model it was possible to find the optimum welding parameters with only 15% of the tests in comparison to a common stepwise investigation. The ultrasonic welding process leads to the displacement of the polymer matrix out of the welding zone and enables a direct contact between the load bearing carbon fibers and the sheet metal without any damage of the carbon fiber reinforcement. Tensile shear strengths of more than 30 MPa were determined for suitable process parameters for AA5754/CF-PA66-joints. The application of off-peak staged welding force profiles enables higher tensile shear strengths of up to 45 MPa. A metal pre-treatment by a corundum blasting process or acid pickling in concentrated nitric acid leads to a further increase to 54 MPa and to a positive effect on the long-term stability of the metal/CFRP-joints. Furthermore, the weldability of the high performance composite CF-PEEK with aluminum sheets was demonstrated. For these joints tensile strengths of up to 40 MPa were realized.

Regarding the efficiency, automation capability, ecological compatibility, achievable mechanical and technological properties ultrasonic metal welding is a very attractive alternative to existing joining techniques for multi-material components. Application fields for this welding method can be seen in the automotive industry, e.g. to join lightweight crash structures or integral body parts by using fiber reinforced materials or in aircraft construction for components of the fuselage or the aerofoil, e.g. for spoilers or blocker doors.

Acknowledgement

The authors would like to thank the German Research Foundation (DFG) for the financial support (Research unit 524).

References

[1]   Wagner, G. et al., *Technical Journal of welding and allied processes*, Issue 5/03 (2003), 268-274
[2]   Rotheiser, J., *Joining of Plastics* (Munich: Carl Hanser Verlag, 2004),
[3]   Krüger, S., Wagner, G., Eifler, D., *Advanced Engineering Materials*, 2004, 6, No. 3: 157-160
[4]   Balle, F., Wagner, G., Eifler, D., *Materials Science and Engineering Technology*, 2007, 38, No. 11: 934-938
[5]   Hill, Th., Lewicki, P., *Statistics - Methods and Applications*, StatSoft, Inc., 2005
[6]   Ehrenstein, G. W., Riedel, G., Trawiel P., *Thermal Analysis of Plastics* (Munich, Vienna: Carl Hanser Verlag, 2004)
[7]   Balle, F. et al., *Materials Testing MP*, 2008, 50, No.4: 184-189
[8]   Balle, F., Wagner, G., Eifler, D., *Advanced Engineering Materials*, 2008, in press

Aluminum Alloys: Fabrication, Characterization and Applications II
Edited by: Weimin Yin, Subodh K. Das, and Zhengdong Long
TMS (The Minerals, Metals & Materials Society), 2009

EFFECT OF PARTICLE LOADING ON THE PROPERTIES OF Al/SiC METAL MATRIX COMPOSITES

Brian Givens, W. Michael Waggoner, Ken Kremer, Michael K. Aghajanian

M Cubed Technologies; 1 Tralee Industrial Park, Newark, DE 19711, USA

Keywords: Metal Matrix Composites, Al/SiC, Mechanical Properties

Abstract

Al/SiC metal matrix composite (MMC) castings offer a significant advantage over traditional unreinforced Al castings. Relative to Al alloys, these composite materials have higher stiffness with only a minor increase in density, and lower coefficient of thermal expansion (CTE). By manipulating the processing of these composites, properties can be tailored for specific applications. Factors that can be varied include: alloy chemistry, reinforcement shape, reinforcement loading, processing method and heat treatment. This study measures the effect of particle loading on cast Al/SiC MMCs. Composites with reinforcement loadings of 0, 30 and 55 volume percent were fabricated. The composites were then characterized for microstructure, composition and properties (density, Young's modulus, tensile strength, fracture toughness, thermal conductivity and CTE).

Introduction

Despite its many advantageous properties of low density, high strength and high toughness, Al has low stiffness (Young's modulus) and high coefficient of thermal expansion (CTE). These deficiencies limit the utility of Al in applications where precision and/or stability are required. Silicon carbide, on the other hand, provides very high stiffness and very low CTE. Thus, incorporation of SiC particles in an Al alloy matrix can allow the shortcomings to be overcome. For comparison, various properties of Al and SiC are provided in Table I.

Table I: Properties of Constituents in Composites

	Aluminum [1]	SiC [2]
Density (g/cc)	2.7	3.2
Young's Modulus (GPa)	70	460
Poisson's Ratio	.29	.14
Coef. Thermal Exp., 20 to 100°C (ppm/K)	24	3.0
Thermal Conductivity (W/mK)	222	300

In addition to improved properties when compared to unreinforced Al, Al/SiC metal matrix composites offer the ability to tailor properties to suit various applications by manipulating the processing of the composites. In order to utilize these materials in actual applications, knowledge of their properties under various types of processing is required. The present paper examines the effect of reinforcement amount on the properties of Al/SiC composites.

Materials

The effect of particle loading was evaluated by producing composites with 0, 30 and 55 volume percent reinforcement. The metal phase of nominally Al-9Si-0.4Mg (similar to A359.1) was used for all compositions. The unreinforced and 30% reinforced samples were cast using traditional foundry methods, using a 30 ppi filter to remove contaminants and dross. At 55% particle loading, the molten Al/SiC material was too viscous to cast in a traditional manner. Thus, energy was applied to form the casting.

The SiC particle size in both the Al/SiC-30p and Al/SiC-55p samples was nominally 23 μm. After fabrication, SiC contents were confirmed using quantitative image analysis (Clemex Vision PE System).

Microstructures of the three material conditions are provided in Figure 1. The unreinforced alloy shows a typical as-cast microstructure of a complex Al alloy, with coarse Al grains and large precipitates at the grain boundaries. The Al/SiC microstructures show SiC particles incorporated into the Al alloy matrix. In the Al/SiC-30p composite, some particle pushing upon solidification is evident, with the particles aligning along the boundaries of the Al grains. Due to very high particle packing in the Al/SiC-55p composite, no grain structure within the Al phase is evident. In both Al/SiC microstructures, the presence of the SiC particles greatly refined the second phase precipitates.

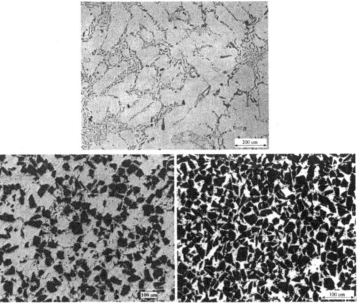

Figure 1. Optical Photomicrographs of Al Alloy (top), Al/SiC-30p (bottom left) and Al/SiC-55p (bottom right), All in As-Cast Condition

Test Methods

The density, elastic properties, tensile strength, fracture toughness, thermal conductivity and coefficient of thermal expansion were determined. Density was measured by the water displacement method per ASTM C135. Elastic properties (Young's modulus and Poisson's ratio) were measured by the ultrasonic pulse echo technique in accordance with ASTM E494. Tensile properties were measured per ASTM B557. Fracture toughness was measured by the 4-point bend chevron notch technique following ASTM C1421. For the fracture toughness property measurements, a Sintech universal testing frame was utilized in conjunction with TestWorks materials testing software. CTE was measured using dilatometry per ASTM E831. Thermal conductivity was calculated using Equation 1 below:

$$\lambda = \kappa * Cp * \rho \tag{1}$$

where λ is the thermal conductivity, κ is the diffusivity, Cp is the specific heat, and ρ is the density. The diffusivity was measured by the laser flash technique in accordance with ASTM E1461. The specific heat was measured by differential scanning calorimetry. The microstructures of the samples were compared using a Leica D 2500 M optical microscope and the Clemex Vision PE imaging software.

Results

Measured properties of the three materials are provided in Table II. All specimens were in the T6 condition. As predicted, the addition of SiC particles to the Al alloy matrix leads to a significant increase in stiffness and a significant decrease in CTE, thus leading to composites with far greater stability than unreinforced Al. The other advantageous changes were small systematic increases in strength and thermal conductivity as the SiC reinforcement was added. On the negative side, the presence of SiC particles reduced fracture toughness and greatly reduced elongation.

Table II: Measured Properties of Al Alloy, Al/SiC-30p and Al/SiC-55p

	Al Alloy	Al/SiC-30p	Al/SiC-55p
Density (g/cc)	2.68	2.78	2.96
Young's Modulus (GPa)	71	120	202
Poisson's Ratio	0.29	0.28	0.25
Ultimate Tensile Strength (MPa)	312	317	340
Elongation (%)	4.6	0.5	0.3
Fracture Toughness (MPa-m$^{1/2}$)	19	15	12
Coef. Thermal Exp., 20 to 100°C (ppm/K)	21.2	15.6	11.8
Thermal Conductivity (W/mK)	134	148	160

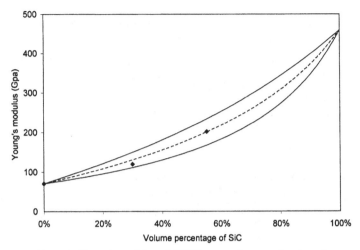

Figure 2: Young's modulus of Al/SiC composites as a function of SiC content. ($\bullet$) The measured values, ($-$) the Hashin-Shtrikman predictive bounds and (---) the average of the upper and lower bounds.

Figure 2 plots the results of Young's modulus versus the volume percentage of SiC. Also, the Hashin-Shtrikman bounds and their average are plotted. The Hashin-Strikman bounds were calculated using the equations in the form provided by Willis [3]:

$$\tilde{\kappa} = \left\{ \sum_i \frac{c_i}{3\kappa + 4\mu} \right\}^{-1} \sum_j \frac{c_j \kappa_j}{3\kappa_j + 4\mu} \qquad (2)$$

and

$$\tilde{\mu} = \left\{ \sum_i \frac{c_i}{6\mu_i(\kappa + 2\mu) + \mu(9\kappa + 8\mu)} \right\}^{-1} \sum_j \frac{c_j \mu_j}{6\mu_j(\kappa + 2\mu) + \mu(9\kappa + 8\mu)} \qquad (3)$$

where c_i is the volume fraction of the ith phase, and κ_i and μ_i are the bulk and shear moduli, respectively, of the ith phase. When κ and μ are chosen such that:

$$\kappa = \max\{\kappa_i\}, \quad \mu = \max\{\mu_i\} \qquad (4)$$

the estimates $\tilde{\kappa}$ and $\tilde{\mu}$ are upper bounds, whereas choosing:

$$\kappa = \min\{\kappa_i\}, \quad \mu = \min\{\mu_i\} \qquad (5)$$

200

results in $\tilde{\kappa}$ and $\tilde{\mu}$ being lower bounds. Bounds on Young's modulus, $\tilde{E}$, can be found from the relation:

$$\tilde{E} = \frac{9\tilde{\kappa}\tilde{\mu}}{3\tilde{\kappa} + \tilde{\mu}} \qquad (6)$$

The elastic properties of the matrix and reinforcement (provided in Table I) were used as input to Equations 2 and 3.

The Hashin-Shtrikman bounds are known to be optimal for the case of a binary composite. Where the properties of a particular composite fall between these bounds depends on the relative compliance of the two phases. If the matrix material is more compliant than the reinforcement, the composites elastic moduli will lie on the lower bound. If the reinforcement is more compliant than the matrix material, then the composite's elastic moduli will lie on the upper bound [4, 5].

In the Al/SiC composites considered here, the matrix material is more compliant than the reinforcement. As the reinforcement amount is decreased, the elastic properties of the composite should approach the Hashin-Shtrikman lower bound. As Figure 2 shows, the data follows this trend.

Conclusions

A series of particulate-reinforced Al/SiC metal matrix composites were produced and characterized. The variable considered was the volume percentage of reinforcement in the composite. Increases in reinforcement amount resulted in significant increases in stiffness and decreases in CTE, as well as small systemic increases in strength and thermal conductivity. At the highest loading (55%), the stiffness was increased by a factor of 2.8 over the unreinforced alloy. As expected, Young's modulus falls near the lower bounds of the Hashin-Shtrikman bounds at the lower reinforcement loading. With the increase in reinforcement amount, two deleterious effects were observed. Elongation was significantly reduced and fracture toughness showed a small decrease.

References

1.   Metals Handbook: Desk Edition (ASM International, Metals Park, OH, 1985)

2.   Engineered Material Handbook, Vol 4, Ceramics and Glasses (ASM International, Metals Park, OH, 1991)

3.   J. R. Willis in "Mechanics of solids, Rodney Hill 60th anniversary Vol.", edited by H. G. Hopkins and M. J. Sewell (Pergamon Press, Oxford, 1982) pp. 237-54.

4.   Z. Hashin, S. Shtrikman, "A Variational Approach to the Theory of the Elastic Behaviour of Multiphase Materials," *J. Mech. Phys. Solids*, **11** 127-140 (1963).

5.   M. K. Aghajanian, R. A. Langensiepen, M. A. Rocazella, J. T. Leighton, C. A. Anderson, "The Effect of Particulate Loading on the Mechanical Behaviour of Al₂O₃/Al Metal Matrix Composites," *J. Mat. Sci.*, **28** 6683-6690 (1993).

Aluminum Alloys: Fabrication, Characterization and Applications II
Edited by: Weimin Yin, Subodh K. Das, and Zhengdong Long
TMS (The Minerals, Metals & Materials Society), 2009

IMPACT PROPERTIES AND MICROSTRUCTURAL EVOLUTION OF WELDABLE AND UNWELDABLE AL-SC ALLOYS

Woei-Shyan Lee and Tao-Hsing Chen

Department of Mechanical Engineering, National Cheng Kung University, Tainan, Taiwan

Keywords: Al-Sc alloy, Impact properties, Strain rate effect, Dislocation

Abstract

This study employs a compressive split-Hopkinson pressure bar to investigate the impact properties of two weldable and unweldable Al-Sc alloys at strain rates ranging from 1.2×10^3 s^{-1} to 5.9×10^3 s^{-1} and temperatures of -100°C, 25°C and 300°C, respectively. The results indicate that for both alloys, the impact properties are found to be significantly dependent on both the strain rate and temperature. Moreover, the flow stress, work hardening coefficient and strain rate sensitivity are higher in the unweldable Al-Sc alloy than in the weldable alloy. The TEM observations reveal that in both alloys, the dislocation density increases with increasing strain rate, but decreasing with increasing temperature. The dislocation density of the unweldable Al-Sc alloy is higher than that of the weldable Al-Sc alloy.

Introduction

Aluminum alloyed with scandium has excellent mechanical properties such as high ductility and formability, super fatigue strength and relatively high temperature resistance, due to the presence of coherent, nanometer size Al_3Sc precipitates [1-4]. Researchers have also shown that the addition of copper (Cu) to Al alloys improves their stress corrosion resistance [5] and enhances their strength and toughness properties. However, Al-Cu alloys are highly susceptible to hot cracking, and are therefore unweldable. Nevertheless, recent studies have shown that the addition of small amounts of Sc to Al-Cu alloys suppresses this hot cracking effect, and thus Al-Sc-Cu alloys continue to attract considerable interest for a range of applications; primarily in the aerospace industry [6].

Structural components are invariably subjected to high strain rate loading conditions during their fabrication or service lives. Moreover, the components used in aerospace, maritime and cryogenic applications are commonly exposed to a wide range of temperatures. The dynamic deformation characteristics of many engineering materials are strongly influenced by the strain rate and temperature [7, 8]. Therefore, it is essential to clarify the effects of strain rate and temperature on the deformation and failure characteristics of these materials in order to ensure that structural components fulfill their design objectives when implemented in real-world environments.

Although the literature contains many investigations into the mechanical properties of weldable Al-Sc alloys and unweldable Al-Sc-Cu alloys, very little information is available regarding the effects of strain rate and temperature on the dynamic mechanical responses of these two alloys. Accordingly, the present study utilises a compressive split-Hopkinson pressure bar (SHPB) to examine the dynamic behaviour of weldable and unweldable Al-Sc

203

alloys impacted at strain rates ranging from 1.2 $\times 10^3$ s^{-1} to 5.9 $\times 10^3$ s^{-1} at temperatures of

-100°C, 25°C and 300°C, respectively.

Material preparation and experimental procedure

The weldable and unweldable Al-Sc alloys used in the present study were supplied in bar form by Taiwan Hodaka Technology Co. Ltd. The mass chemical compositions of the two alloys are shown in Table 1. To ensure an optimum dispersion of the Al$_3$Sc precipitates in the matrix, the extruded bars were solution heat treated in an air furnace at 465°C for 1 h, water quenched at room temperature, and then aged naturally for 3 days. The bars were then aged artificially in a two-step heat treatment process performed at 105°C for 7 h and 150°C for 10

h, respectively. Finally, test specimens with a length of 7±0.1 mm and a diameter of 7.2 mm were machined from the heat-treated extruded bars using a centre-grinding process.

The dynamic impact tests were performed at temperatures of −100°C, 25°C and 300°C, respectively, using the compressive SHPB system [8]. The weldable Al-Sc alloy specimens were tested at strain rates of 1.3 $\times 10^3$ s^{-1}, 3.6×10^3 s^{-1} and 5.9×10^3 s^{-1}, respectively, while the

unweldable Al-Sc alloys were impacted at strain rates of 1.2 $\times 10^3$ s^{-1}, 3.2×10^3 s^{-1} and 5.8×10^3

s^{-1}, respectively. TEM specimens were prepared by cutting foils with a thickness of 350 μm from the deformed specimens. Disks with a diameter of 3 mm and a thickness of 150~200 μm , were in a solution of 30% nitric acid and 70% methanol at a temperature of -30

°Cusing an agitation voltage of 15 V.

Table 1 Chemical composition of weldable and unweldable Al-Sc alloys.

Elements (wt.%)	Zn	Mg	Cu	Sc	Zr	Al
Unweldable	8.9	1.7	2.1	0.13	0.15	balance
Weldable	6.8	2.2	--	0.13	0.15	balance

Results and discussion

Figure 1(a) presents the stress-strain curves for the weldable and unweldable Al-Sc alloys deformed at a temperature of −100°C. It can be seen that in both cases, the flow stress increases with increasing strain rate and increasing strain. It is also observed that for a given strain rate and strain, the unweldable Al-Sc alloy has a higher flow stress than the weldable Al-Sc alloy. Figures 1(b) and 1(c) present the stress-strain curves of the two alloys at deformation temperatures of 25°Cand 300°C, respectively. The results show that for a given strain rate and strain, the flow stress of the unweldable Al-Sc alloy is higher than that of the weldable Al-Sc alloy at both values of the deformation temperature. Overall, Figures 1(a)~1(c) reveal that the fracture strain decreases as the strain rate increases or the temperature decreases. However, it is noted that the specimens deformed at room temperature under a strain rate of 1.2×10^3 s^{-1} (unweldable Al-Sc alloy) or 1.3×10^3 s^{-1} (weldable Al-Sc alloy), or at a temperature of 300°C and any of the current strain rates, do not fracture, even when deformed to true strains of 0.48 (unweldable Al-Sc alloy) or 0.60 (weldable Al-Sc alloy). Figure 1(d) superimposes the stress-strain curves of the two alloys when deformed at a strain rate of approximately 3×10^3 s^{-1} and temperatures of -100°C, 25°C and 300°C, respectively. The results clearly show that for both alloys, the flow stress decreases with increasing temperature.

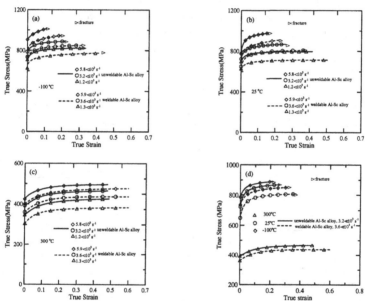

Fig. 1: True stress-strain curves of unweldable Al-Sc alloy deformed at strain rates of 1.2×10^3 s^{-1}, 3.2×10^3 s^{-1} and 5.8×10^3 s^{-1} and weldable Al-Sc alloy deformed at strain rates of 1.3×10^3 s^{-1}, 3.6×10^3 s^{-1} and 5.9×10^3 s^{-1} at temperatures of: (a)–100°C; (b) 25°C; (c) 300°C. (d) True stress-strain curves of weldable and unweldable Al-Sc alloys deformed at approximately equal strain rates at temperatures of –100°C, 25°C and 300°C, respectively.

The stress-strain behaviours of the weldable and unweldable Al-Sc alloys can be described by the following empirical power law of $\sigma_t = \sigma_y + B\varepsilon_t^n$, where σ_y is the yield strength, B is the material constant, and n is the work hardening coefficient. Table 2 presents the values of σ_y, B and n obtained by substituting the experimental data plotted in Figures 1(a)~1(c) into this simple power law. It can be seen that for both alloys, the yield strength, material constant and work hardening coefficient all increase with increasing strain rate, but reduce with increasing temperature. Furthermore, it is apparent that for a given strain rate and temperature, the unweldable Al-Sc alloy has a higher yield strength, material constant and work hardening coefficient than the weldable Al-Sc alloy.

Figure 2 plots the flow stress against the logarithmic strain rate at a true strain of 0.1 and temperatures of –100°C, 25°C and 300°C, respectively. It is evident that both alloys exhibit a strain rate sensitive behaviour and that the flow stress decreases with increasing temperature at a constant strain rate. Moreover, it is observed that for a similar strain rate, the flow stress induced in the unweldable Al-Sc alloy is higher than that induced in the weldable alloy. The strong dependence of the flow stress on the temperature and logarithmic strain rate suggests that the deformation of both alloys is dominated by a thermally-activated mechanism. The experimental data plotted in Figures 1(a)~1(c) can be used to calculate the strain rate sensitivity, β, of the two Al-Sc alloys in accordance with $\beta = (\partial\sigma/\partial\ln\dot\varepsilon) = (\sigma_2 - \sigma_1)/\ln(\dot\varepsilon_2/\dot\varepsilon_1)$ where the flow stresses σ_1 and σ_2 are obtained from tests conducted at average strain rates of $\dot\varepsilon_1$ and $\dot\varepsilon_2$, respectively, and are calculated at the same value of plastic strain. The corresponding results are summarised in Table 3. For both alloys, it can be seen that for a constant strain, β increases with increasing strain rate, but decreases with increasing temperature. Furthermore, for a constant strain rate and temperature, β increases with increasing strain. Moreover, for a given strain rate, strain and temperature, the strain rate sensitivity of the unweldable Al-Sc alloy is higher than that of the weldable Al-Sc alloy.

Table 2 Dynamic mechanical properties of weldable and unweldable Al-Sc alloys impacted at different strain rates and at temperatures.

	$\dot\varepsilon\,(s^{-1})$	T (°C)	σ_y MPa	B (MPa)	n	ε_f	σ_f (MPa)
		300	314	101	0.16	--	--
	1.2×10^3	25	706	111	0.19	--	--
		-100	732	122	0.23	0.33	823
Unweldable Al-Sc alloy		300	365	109	0.18	--	--
	3.2×10^3	25	724	218	0.27	0.26	867
		-100	750	226	0.28	0.22	890
		300	394	124	0.21	--	--
	5.8×10^3	25	783	369	0.32	0.16	976
		-100	829	486	0.47	0.11	1014
		300	253	137.7	0.15	--	--
	1.3×10^3	25	583	153.5	0.17	--	--
		-100	602	207	0.2	0.42	774.3
Weldable Al-Sc alloy		300	318	138.3	0.17	--	--
	3.6×10^3	25	606.5	254.2	0.2	0.37	806.4
		-100	697	209	0.24	0.32	849.2
		300	365	140.2	0.20	--	--
	5.9×10^3	25	644.3	412.5	0.28	0.23	906.4
		-100	780	283	0.32	0.21	930.3

Figures 3(a)~3(c) present the dislocation structures of unweldable Al-Sc specimens deformed at temperatures of -100 °C and $1.2\times10^3s^{-1}$, -100 °C and $5.8\times10^3s^{-1}$, and

300°C and $5.8\times10^3\text{s}^{-1}$, respectively. Figures 3(d)~3(f) present the dislocation structures of weldable Al-Sc alloy specimens deformed at -100°C and $1.3\times10^3\text{s}^{-1}$, -100°C and $5.9\times10^3\text{s}^{-1}$, and 300°C and $5.9\times10^3\text{s}^{-1}$, respectively. Comparing these images with those shown in Figures 3(a)~3(c), it is evident that the dislocation density increases with increasing strain rate, but decreases with increasing temperature. It is also found that the dislocation density of the unweldable Al-Sc alloy is higher than that of the weldable Al-Sc alloy. A higher dislocation density increases the degree of dislocation tangling, and therefore reduces the mobility of the dislocations. As a result, the resistance of the unweldable Al-Sc alloy to plastic deformation is enhanced.

Figure 4(a) shows that for both alloys, the dislocation density increases with increasing strain rate, but decreases with increasing temperature. Figure 5(b) shows the relationship between the reciprocal of the cell size and the dislocation density as a function of the temperature. Note that the results correspond to a constant true strain of 0.1 for the unweldable Al-Sc alloy and 0.2 for the weldable Al-Sc alloy. It can be seen that the dislocation cell size decreases with an increasing dislocation density. Furthermore, the dislocation density reduces with increasing temperature.

Table 3 Strain rate sensitivity of weldable and unweldable Al-Sc alloys as function of strain rate, true strain and temperature.

Unweldable Al-Sc alloy	Temperature °C	Strain rate range $1.2\times10^3\text{ s}^{-1} \sim 3.2\times10^3\text{ s}^{-1}$ ε_t			Strain rate range $3.2\times10^3\text{ s}^{-1} \sim 5.8\times10^3\text{ s}^{-1}$ ε_t		
		0.05	0.1	0.15	0.05	0.1	0.15
Strain rate sensitivity β (MPa)	300	42.5	42.6	43	57.3	59.3	59.7
	25	61.3	66.8	71.0	191.2	195.8	199. 5
	-100	70.4			218.2		
Weldable Al-Sc alloy	Temperature °C	Strain rate range $1.3\times10^3\text{ s}^{-1} \sim 3.6\times10^3\text{ s}^{-1}$ ε_t			Strain rate range $3.6\times10^3\text{ s}^{-1} \sim 5.9\times10^3\text{ s}^{-1}$ ε_t		
		0.05	0.1	0.15	0.05	0.1	0.15
Strain rate sensitivity β (MPa)	300	35.6			45.2		.7
	25	55.7	.6		148.2	155.1	166.
	-100		66.4	68.2	166.1		189.1

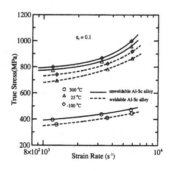

Fig. 2: Variation of true stress with log strain rate for weldable and unweldable Al-Sc alloys at true strain of 0.1 and temperatures of −100°C, 25°C and 300°C.

Conclusion

The results have shown that for both alloys, the flow stress increases with increasing strain rate, but decreases with increasing temperature. Moreover, it has been found that the unweldable Al-Sc alloy is consistently stronger than the weldable alloy. For both alloys, the strain rate sensitivity increases with increasing strain rate, but decreases with increasing temperature. The TEM observations have revealed that for both alloys, the dislocation density increases with increasing strain rate, but decreases with increasing temperature. Furthermore, the dislocation density of the unweldable Al-Sc alloy is higher than that of the weldable Al-Sc alloy. Thus, the dislocation cell size of unweldable Al-Sc alloy is lower than that of weldable Al-Sc alloy.

Acknowledgement

The authors gratefully acknowledge the financial support provided to this study by the National Science Council (NSC) of Taiwan under contract no. NSC 96-2221-E-006-048.

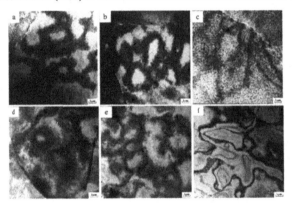

Fig. 3: TEM micrographs of dislocation microstructure of unweldable Al-Sc alloy deformed at (a) -100°C and $1.2×10^3$ s^{-1}, (b) -100°C and $5.8×10^3$ s^{-1}; (c) 300°C and $5.8×10^3$ s^{-1}; Weldable Al-Sc alloy deformed at (d) -100 °C and $1.3×10^3$ s^{-1}, (e) -100°C and $5.9×10^3$ s^{-1}; (f) 300°C and $5.9×10^3$ s^{-1}

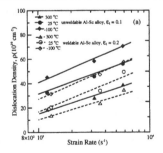

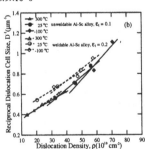

Fig. 4: (a) Variation of dislocation density with strain rate at temperatures of −100°C, 25°C and 300°C for constant strain of 0.1 for unweldable Al-Sc alloy and 0.2 for weldable Al-Sc alloy; (b) Variation of reciprocal of dislocation cell size with dislocation density at temperatures of −100°C, 25°C and 300°C for constant strain of 0.1 for unweldable Al-Sc alloy and 0.2 for weldable Al-Sc alloy.

REFERENCES

1. I. Kentaro and M. Yasuhiro, "Dynamic recrystallization in Al–Mg–Sc alloys," *Mater Sci Eng A*, 387-389 (2004), 647-650.
2. R. R. Sawtell and C. L. Jensen, "Mechanicla properties and microstructures of Al-Mg-Sc alloys," *Metall Trans A*, 21A (1990), 421-430
3. V. Jindal, P.K. De and K. Venkateswarlu, "Effect of Al₃Sc precipitates on the work hardening behavior of aluminum–scandium alloys," *Mater Let*, 60 (2006), 3373-3375.
4. T. Torma et al., "Hardening mechanism in Al-Sc alloys," *J Mater Sci*, 24 (1989), 3924-3927
5. V. V. Zakharov, T. D. Rostova and I. A. Fisenko, "High-Strength Weldable Corrosion-Resistant Aluminum Alloy for Bearing Building Structures," *Met. Sci. Heat Treat*, 47 (2005), 377-382.
6. Z. Ahmad, "The properties and application of scandium-reinforced aluminum" *JOM*, 55 (2003), 35-39.
7. Y.B. Xu et al., "Shear localization and recrystallization in dynamic deformation of 8090 Al–Li alloy," *Mater Sci Eng A*, 299 (2001), 287-295.
8. W.S. Lee and C.F. Lin, "Plastic deformation and fracture behaviour of Ti–6Al–4V alloy loaded with high strain rate under various temperatures," *Mater Sci Eng A*, 241 (1998), 48-59.

AUTHOR INDEX
Alumiunum Alloys: Fabrication, Characterization and Applications II

SUBJECT INDEX
Alumiunum Alloys: Fabrication, Characterization and Applications II